THINKING
ABOUT
CREATION

THINKING
ABOUT
CREATION

ETERNAL TORAH
AND MODERN PHYSICS

ANDREW GOLDFINGER

JASON ARONSON INC.
Northvale, New Jersey
Jerusalem

This book was set in 11 pt. Galliard by Hightech Data Inc., of Bangalore, India, and printed and bound by Book-mart Press, Inc. of North Bergen, NJ.

Library of Congress Cataloging-in-Publication Data

Goldfinger, Andrew.
    Thinking about creation : eternal Torah and modern physics / by Andrew Goldfinger.
       p. cm.
    Includes index.
    ISBN 0-7657-6042-8
    1. Creation. 2. Creation—Religious aspects—Judaism.
3. Evolution (Biology) 4. Evolution (Biology)—Religious aspects—
Judaism. 5. Judaism and science. I. Title.
BS651.B76   1998
296.3'4—dc21   98-40593

Printed in the United States of America on acid-free paper. For information and catalog write to Jason Aronson Inc., 230 Livingston Street, Northvale, NJ 07647-1726, or visit our website: www.aronson.com

In memory of my parents,
Tzvi Hirsch ben Avraham David
Yucheved bas Moshe

Contents

Contents

Acknowledgments

I want to thank some special people who contributed to this work:

First, I would like to thank the Bostoner Rebbe, Rabbi Levi Yitzhok Horowitz, for giving me my direction in Torah, for years of wise counsel, and for encouraging me to marry my wife. I would like to thank Rabbi Chaim Horowitz for turning me on to Torah in the first place.

Thanks are in order to a number of people who read early drafts of this work and provided many helpful comments. In particular, I am grateful to Harold Gans, Mitch Persons, Dovid Shulman, and Rabbi Yerachmiel Fried. I am also grateful to Thelma Blumberg for reading sections of the revised text.

I would like to thank Rabbi Noah Weinberg for his tireless efforts in leading, inspiring, and cajoling those of us who have been given the privilege of being part of Aish HaTorah and the Discovery Seminars, and Rabbi Eric Coopersmith for the many hours he has spent on coaching, quality control, and hand holding. Truly, the weekends I have spent lecturing at

Discovery seminars have been among the most significant and precious of my life.

Thank you to Debbie Howarth for her help in explicating the "seemingly irrelevant" material in the Joseph story, and for many long hours on the telephone tracking down hard to find sources.

I am grateful to Rabbi Tzvi Feldman, Dr. Edward Tryon, Academic Press, Maaliot Press, Mesorah Publications, and the American Institute of Physics for their kind permission to reprint quoted material. The definition in chapter 3 is quoted by permission, from *Merriam Webster's Colligiate ® Dictionary, Tenth Edition* © 1997 by Merriem-Webster, Incorporated. The quote in Chapter 11 is from *The New Columbia Encyclopedia*, ed. William Harris and Judith Levy © 1975 Columbia University Press. Reprinted with the permission of the publisher.

My dear friend Harold Gans has been an inspiration and a sounding board. His contributions fill these pages. The book could not have existed without him.

Finally, I would like to thank my wife, Shana Goldfinger. Of course, without her help, there would be no book. But also, in a very true sense, neither would there be me.

Preface

I love Torah, and I also love science. When I look at them together, I cannot resist speculating.

For example, there are two food substances that have special places in Jewish law: bread and wine. Bread represents all food, so when we "establish a meal" by pronouncing a special blessing on bread, we do not have to pronounce separate blessings on the other foods that follow. The same is true of wine and the beverages that follow it. Both bread and wine come in superior and lesser forms. The most desirable bread is leavened, but the lesser form, unleavened matzoh, is also bread under Jewish law. Similarly alcoholic wine is of the highest quality, but unfermented grape juice can also be used when wine is called for.

Both bread and wine are taken from their lesser to their superior forms by the action of yeast. Yeast acts on sugar and produces two products: alcohol and carbon dioxide. It is the gaseous carbon dioxide that causes bread to rise, and the alcohol is driven off during the baking. With wine, it is

xi

the alcohol we want, and we allow the carbon dioxide to escape. Yeast has exactly two products, and each of them is closely involved in the improvement of one of the special foods. Why?

Another speculation: there are two principal names of God. One has four letters. It indicates God as the source of mercy. The second has five letters and indicates justice. There is a tension between justice and mercy, and this same tension is found in the structure of the Passover seder in which the numbers four and five abound (four cups we drink, a fifth awaits the Prophet Elijah; a fourfold versus fivefold magnification of each plague; and so on). The fundamental molecules of life are called DNA and RNA. Each is made up of four elemental structures called bases. DNA has four bases, and RNA has four, but although they share three in common, the fourth bases of each differ. Thus there are altogether five bases, of which RNA and DNA each contain four. Is the four-five dichotomy, the tension between justice and mercy, found at the very foundations of life itself?

What does all this mean? I don't know. But it cannot be accidental. God created an exciting world in which His presence is manifest. I have only to look into the skies at night, or into the intricacies of molecular biology to see structures of incredible beauty. It is a world in which puzzles abound, and I delight in tackling them. The greatest of the puzzles is the nature of the creation, and I truly enjoy speculating about this. I enjoy it so much, that I have written this book as a sort of progress report.

I hope that the reader will not get the idea that I understand how it all fits together. God has created a world that abounds in puzzles, and we can only understand it approximately. But I am glad for this. It gives me an inexhaustible source of puzzlement and beauty, and I hope that the reader will share my enthusiasm for the thought process as he turns the pages below.

Tam V'Nishlam Shevach L'Kel Borai Olam
Andrew D. Goldfinger

1

July 10, 1925
The Problem

In the summer of 1925, in the town of Dayton, Tennessee, a duel began between two respected men. The battle was to last eleven days and would capture the attention of the continent.

At issue was the behavior of a young man named John Thomas Scopes, a teacher of science at Rhea High School. The laws of the state of Tennessee forbade the teaching of any ideas that threatened the biblical belief that man had been created by God. As a biologist, John Scopes taught his class the Darwinian theory of evolution: that all living things had slowly and accidentally evolved from simpler organisms and that man was descended from the apes. The young teacher had broken the law. The prosecution of him in court was to be forever known as the "Scopes Monkey Trial."

The duel took place in the courtroom between William Jennings Bryan, who led the prosecution, and Clarence Darrow, the principal attorney for the defense. William Jennings Bryan was a prominent national figure. He had served in Congress, run twice for president, and was a charismatic orator. He was also a religious man who believed firmly in the literal truth of the Bible. He found himself among friends in the hot Tennessee courtroom. Clarence Darrow was one of the most prominent lawyers of his day. He faced a tough challenge in his defense of Scopes.

In a stroke of genius, Darrow called Bryan as a witness. Bryan was eager to avail himself of the limelight. His cause seemed safe as he ascended the witness stand to defend religious fundamentalism, but he encountered a formidable opponent in Darrow. The defense attorney countered his every argument as he led him through the logical and biological problems in the biblical text. In the end, Bryan was crushed on the stand. He fell from public respect, and with him the entire edifice of literal belief in Genesis was cast into disrespect. The religious literalist was exposed as a backward scientifically illiterate simpleton. Science was seen to be the progressive wave of the future and the only path toward truth. It scarcely mattered that Scopes was actually found guilty and fined one hundred dollars. In the court of American public opinion, religious literalism was overturned and destroyed. Bryan died five days later.

It is no longer 1925. Decades have passed, and the play *Inherit the Wind* has popularized the trial. The movie by the same name is best known for the performances of Frederick March as Bryan and Spencer Tracy as Darrow. Audiences cheer for the good guys, and everyone knows who they are. They delight at the fall of prejudice, ignorance, and Bryan. In the Tennessee of the late twentieth century, the issue is the legality of teaching "creationism" along with evolution, and the religious viewpoint is losing.

And yet, also in the America of today, we see a resurgence of religious belief. We see the massive movement of religious returnees, the flourishing of the yeshivas, the growth of organizations such as the Association of Orthodox Jewish Scientists. We know that these are all modern people. We know that many of them are scientifically knowledgeable. Yet, we also see that they actually take the Bible seriously and Genesis to be literally true. How can such a thing be possible?

How can a modern believing Jew deal with this? Does the scientific viewpoint contradict the biblical account of creation, as we have seen in the movie, or does it not? Was the world created in six days, or did it take billions of years? Did man spring into being fully formed, or did he evolve from the apes? What is the truth? Can we tell what really happened? What path should we take as we think about creation?

2

Bumblebees
The Conclusion

Let us begin our discussion with its conclusion. This may seem a bit odd, but we are not after drama; we are after understanding.

Wouldn't it be nice to show that there is absolutely no conflict between the scientific and Torah viewpoints of creation, that they are completely consistent in all their details? Would that we could do this, but unfortunately we cannot. Yes, we will see that there are striking parallels between both viewpoints, but in the end a few apparent contradictions will remain.

How are we to deal with the remaining problems?

Let us be honest. There are many intelligent modern people who do not believe that the Torah was given by God. When they see a contradiction with the scientific viewpoint, they are not troubled. They simply say that the Torah is wrong. End of the matter.

There are also many Torah-believing Jews who do not trust science as it is usually practiced. They reject the Darwin-

5

ian theory of evolution as a lot of unproven, if not downright dangerous, nonsense. They consider evolutionary biologists to be atheists who, from the start, are biased in their approach to nature. We will examine this position in greater detail later, but for now let us observe that people who take this view face no problem in reconciling science and Torah. They simply reject the former when contradicted by the latter.

But what of those of us who regard both disciplines as valid? What about those of us who find much to admire in the objectivity of science yet firmly believe that the Torah was given to Moses letter by letter by God and that all within it is literally true?

Perhaps we can sympathize with the poor aeronautical engineers of fifty years ago. At this time, engineers knew how to design airplanes that flew pretty well. To do this, they employed the theories of aerodynamics, that blend of physics and applied mathematics that describes the motions of gasses such as air. We can imagine how fascinating it was to branch out and apply these proven theories to new domains, to how birds fly and how sailboats harness the wind. We can also picture the amazement of the first engineer to perform a theoretical study of the bumblebee.[1]

The problem was this: according to the well-established and tested theories of aerodynamics, the bumblebee could not

---

[1]This information about the bumblebee comes under the category of a "well-known story." It was well enough known to motivate the staff of the laboratory in which I work to jocularly name a project "the bumblebee project" since critics had said it would never fly. Although I have been unable to locate a reference to the original calculations in the aerodynamic literature, I have spoken with Dr. Alvyn Eaton of the Johns Hopkins University/Applied Physics Laboratory, who performed the follow-up calculations using moving wings that showed that the bee could, indeed, fly.

fly. Nobody told the bumblebee about this, so it continued its usual pastime of gathering nectar, pollinating plants, and—flying. There was a contradiction between two widely trusted means of obtaining knowledge: scientific analysis and direct observation. (After all, we have all *seen* bumblebees fly.)

How did the academic world react? Many had a good laugh. Those who were suspicious of science, who were even more suspicious of the practitioners of science, experienced a vicarious victory. The arrogant scientists who thought they could explain the whole world were outsmarted by the little bumblebee.

What of the scientists themselves? They differed from the scoffers in one important way—they had confidence in the methods of their craft. They were not willing to throw away aerodynamics because of an apparent contradiction. They realized that they had to get down to work and figure out what was going on. Far from being crushed, they were intrigued, fascinated, and challenged, for this is the personality of the scientist: curiosity and the thrill of a new problem to be solved. A contradiction is not a threat; it is an opportunity to learn something new.

Let us emulate these scientists. We will welcome difficulties as we face the challenges of comparing the scientific and Torah viewpoints of creation. If the two disciplines seem to come to different conclusions, we could simply throw one of them away. This would be an honest and intelligent thing to do if we had little confidence in one of the approaches. But if we regard both as valid, we have a problem. Then, as the scientist, we do not get upset or depressed. We instead become intrigued and motivated. We are driven to get down to work to try and figure out what is going on.

In the following pages we will assume that both science and Torah study are valid methods of learning about the world. This book is not for readers who are willing to let go of one of them. It is for those of us who are confronted by the problems and feel the thrill of the challenge. We may not be able to solve all the problems, but we can, at least, experience the exhilaration of inquiry as we think about creation.

We will begin by taking a few chapters to set down the ground rules: what science is, and what it is not; what Torah is, and what we can expect to learn from it; how we are to think about contradictions, and how we are to approach the text of the Torah. Only then, in Chapter 7, will we begin to tell the story of creation. We hope that the reader will resist the temptation to look ahead.

(By the way, the scientists finally did work things out with regard to the bumblebee. It turns out that the early calculations assumed that the bee's wings do not flap when it flies. On real bees, they do. At first, scientists thought this detail was unimportant. They were wrong.)

3

What Is and
What Ought to Be
The Limitations of Science

Late at night, a medical researcher—let us call her Dr. Selma Wainwright—is sitting alone in her laboratory. Surrounding her are hundreds of thousands of dollars worth of impressive technical equipment—electronic screens, automated chemical analysis systems, and computer terminals. But the hardware and software are not in use. What is in action are the few ounces of "wetware" within her head (yes, scientists really do use this term). She has an idea.

She has been looking over some recent scientific papers and has noticed that aspirin can interfere with the clotting of blood. She knows that blood clots are a major cause of heart attacks, and she wonders if aspirin might be a useful drug in preventing them. She designs an experiment. She will give aspirin to a group of people who are at risk of coronary disease and see how many of them go on to have heart attacks in the next year. As a careful researcher, she knows that there exists in medicine something known as the placebo effect. That is, a certain number of patients with most any complaint who are given placebos (fake pills that don't contain any drug) will improve. No one really knows why this happens, but all medical researchers are aware of it and must consider it in their experiments. To deal with the placebo effect, Dr. Wainwright will use two groups of patients, both of equal size, both of them at risk of coronary disease. The groups will be matched as closely as possible: the same number of men and women; the same number of smokers and nonsmokers; and the same distribution of ages. One group will be given aspirin, the other group (called the control group) will be given placebos. (We will not discuss the ethical implications of this experiment.)

A year later, she compares the two groups. The patients in the group given aspirin have had significantly fewer heart attacks than those in the control group. Her hypothesis (that aspirin protects against heart attacks) has been confirmed. She publishes the results in a scientific journal. A few weeks later, newspaper headlines scream the news: ASPIRIN PROTECTS AGAINST HEART ATTACKS. Dr. Wainwright sees the headlines, smiles, discards the paper, and sits down to plan doing the same experiment all over again.

Let's consider what has happened. In the opening scene we found Dr. Wainwright alone in her laboratory. All her technical equipment was turned off, yet she was practicing science. We all have a tendency to confuse science and technology. Technological devices are the products of engineering. They are designed using principles established through the scientific method, but they are only tools. Modern technology is useful to scientists, but it is also useful for cooking food and diapering babies.

Science is characterized by inquiry. A scientist asks questions about how the world works and tries to answer these questions through the scientific method. This is the method used by Selma Wainwright.

She began by clearly stating her **hypothesis**: aspirin could protect against heart attack. She then designed an **experiment** to test this hypothesis. It was not sufficient to merely give aspirin to an arbitrary group of people and see what happened. There are so many factors that influence heart attacks that she could not be sure whether it was the aspirin that was having the effect. It was necessary to choose a **control** group, one matched to the experimental group in every way except the one factor being tested. It was even necessary to give the control group placebos in case the psychological effect of taking a pill had some unknown effect. Finally, in her journal article she did not state that she had learned some absolute truth; she stated with care merely that the experiment had **confirmed** her hypothesis. She was amused at the overreaction of the newspaper, for she knew that good scientific practice requires that experiments be **repeatable**. So she put the newspaper away and got back to work.

This last point bears repeating (no pun intended). In the popular press, scientific information is often regarded as truth. Conscientious scientists do not speak of learning the truth. They realize that science is always somewhat tentative. Oh— of course—there are some scientific principles that have been confirmed in such depth that we regard them as almost certainties. After all, most of us are willing to get into airplanes and trust our lives to aeronautical engineers. But theories, no matter how widely held, have been known to collapse suddenly in the face of just one more experiment. Let us recall the poor little bumblebee.

The confusion of scientific theory with truth has been called scientism.[1] It is almost a religious belief, held on faith. In many ways, it is the religion of our times. But the scientific method is imperfect and incomplete. It is not hard to identify pitfalls.

It is easy, in complex situations, to make mistakes in designing controlled experiments. An important factor could be overlooked. A result could be misinterpreted. In some cases, it is purely and simply not possible to perform a true controlled experiment.

Consider the field of astronomy. An astronomer cannot manipulate the stars. He can only observe what is there. Does this mean that he cannot experiment? Not really. An experiment in astronomy is performed by formulating a hypothesis and then looking into the heavens to see if it is confirmed. The astronomer may hypothesize that stars are formed from large clouds of gas. If this were so, he would expect to see many clouds of gas with newly formed stars within them. The astronomer can then carry out an experimental program of looking for gas clouds and counting how many of them contain young stars. This is not a truly controlled experiment, but it is the best he can do.

The same situation applies to evolutionary biology. With only a few exceptions (fruit fly experiments being one of them), experiments cannot be carried out directly. Rather, the biologist must gather fossils, form hypotheses, and then try to predict what he will find in other layers of rock if his theo-

---

[1] Merriam-Webster's Collegiate ® Dictionary, Tenth Edition. (Springfield, MA: Merriam–Webster, 1997) defines scientism to be: "an exaggerated trust in the efficacy of the methods of natural science applied to all areas of investigation (as in philosophy, the social sciences, and the humanities)."

ries are correct. Darwin, for example, predicted that we would find fossil evidence of a gradual change of one species of animal into another. Paleontologists have been carrying out Darwin's experiment for over a hundred years, and we will discuss their results in some detail in a later chapter.

These limitations of astronomy and evolutionary biology do not make for bad science. They merely point out how difficult and tentative even the best science is. There is also a human factor.

How does a scientific paper get published? The author submits it to a journal, and the editor then sends copies of the paper to a group of anonymous experts in the field who are called referees. The referees read the paper, consider how careful and important the work is, and then submit their votes as to whether or not it should be published. This process is called peer review. Although it is practiced by all reputable scientific journals, it is not hard to see flaws in it. Popular theories are likely to be published, while unpopular ones may never make it into print. Yet, it is hard to think of a better system short of simply publishing *all* papers that are submitted (which is economically impossible).

Usually, the peer review system works, but there have been some notable failures. A fascinating case concerns the coelecanth (see'-la-canth) fish. This fish was "known" to have become extinct several million years ago. Therefore, if fossils of it were found in a layer of rock, that rock layer must have been very old. What happened if a biologist submitted a paper in which he found a relatively modern fossil in a layer with a coelecanth fish? Either it would "prove" that the modern fossil was not so modern after all, and this fossil would be redated, or his paper would be rejected by the referees as incorrect. In 1938, a live coelecanth fish was found swimming off the shore of South Africa. Since then others have been

discovered elsewhere.[2] It was not quite as extinct as paleontologists thought.

Well, OK, scientists were willing to admit their mistake,
but what of all those papers that had been rejected by the
referees over the years? What of all the other fossils that were
incorrectly dated? Did anyone go back and try to straighten
it all out? In truth, this would have been an almost impossible task.

What we have described above is not really bad science
or dishonesty. Researchers are human, and they can only do
their best. But the best science has its limitations.

Let us consider a more subtle limitation. We have seen
that scientific practice can sometimes produce conclusions that
are not correct. Is it possible that there are facts that are correct, that are "true," yet cannot be confirmed by the scientific
method?

A number of years ago a woman in Baltimore was get-
ting into her car when she suddenly felt a sense of foreboding about her baby. She left her car, ran upstairs, and found
the infant entangled in the straps of a crib toy, unable to
breath. The baby's life was saved. We call this extrasensory
perception (ESP), and we are all familiar with the debate over
whether it exists. Scientists call a single case such as this an-
ecdotal. By this, they mean that it is interesting and may suggest a hypothesis, but that the event could have been a random coincidence. Attempts to confirm the hypothesis of ESP
through controlled experiments have not been successful. Does
this mean that it does not exist?

What if extrasensory perception occurs only during
emotionally charged, life-threatening circumstances? It would

---

[2]*The New Columbia Encyclopedia*, ed. William H. Harris and Judith
S. Levey (New York: Columbia University Press, 1975), p. 1598.

not be possible (in a moral sense) to conduct controlled experiments to test this hypothesis, yet ESP could still exist in reality.

Let us consider a more astounding limitation of the scientific method. Science must proceed through controlled experiments. The experimental and control groups must be identical in all ways except for the factor being tested. When Dr. Wainwright established her two groups, they were matched in terms of sex, age, and smoking pattern. They were as alike as she could make them, but they still consisted of different people. As far as she was concerned, the fact that one middle-aged man was named Fred and came from Chicago, while the matching man in the control group was named Walter and was raised in Spokane, did not matter. She truly believed these differences to be irrelevant. Is it always possible to identify such irrelevant differences with any certainty?

Torah tells us that we are all unique. It also tells us that all events in the world are interrelated. It is not true that the life of a Jew in Memphis is unaffected by the performance of a good deed by another Jew in Hebron. We have a principle: *kol yisrael arevim zeh lezeh*. Each Jew is responsible for all the others; each Jew is influenced by all others. They are not irrelevant in his life. When it comes to people, at least, totally independent control groups are impossible.

In addition to being controlled, scientific experiments must be repeatable. The results tomorrow must duplicate the results today. But Torah tells us that the world changes as human history progresses towards its conclusion. Today is not the same as yesterday, nor will tomorrow be as today. We might speculate that circumstances exist in which there are truly *no* irrelevant factors, no repeatability.

In the 1930s a shock wave rippled through the world of mathematics. A young mathematician named Kurt Goedel had proved a disturbing theorem.[3] He showed that in any reasonable mathematical system there are facts that are true, but that cannot be proved mathematically. This violated the deeply held assumptions of most mathematicians and showed that there was a fundamental, inescapable limitation to mathematics.

In principle, we see that a similar limitation may exist in the scientific method. ESP may occur. There may exist a real effect upon the human body from the wearing of mixed wool and linen (forbidden by the written Torah).[4] There may be a threat to health due to the simultaneous consumption of meat and fish (forbidden by rabbinic enactment).[5] Eventually, scientific experiments may confirm one or all of these hypotheses. Or they may not. But science cannot ever *prove* that they are false. There is indeed a possibility that they may all be true yet not subject to scientific confirmation. We may have to look to other methods of gaining knowledge to know for sure. This we will discuss further in the next chapter.

Again, we stress that this does not make science a poor way of learning about the world—far from it. But it does point out some significant limitations that we must take into account as we compare science and Torah.

So far we have been discussing the methodology of science, how it reaches conclusions. But what of the subject matter of science? What sort of questions lie within the proper domain of science? Science concerns itself with the nature of the physical world: what it is made of, what laws govern its behavior, how it may be controlled or manipulated. But can it tell us what is right or wrong?

---

[3]A. K. Dewdney, *The Turing Omnibus* (Rockville, MD: Computer Science Press, 1989), chap. 5.
[4]Deuteronomy 22:11
[5]*Shulchan Aruch, Yoreh Deah* 115:2

Consider the issue of stealing. In most societies it is not allowed. But why? Can an understanding of the physical world, a total description of what *is*, enable us to understand why it is wrong to steal? Larry Thomas doesn't know, so he confronts Dr. John Newman, a professor of social science at a leading university. Their conversation goes something like this:

"Professor Newman, from your viewpoint, why is it wrong to steal?"

"Larry, that is very easy to answer. For any society to function, there must be some control over who is allowed to use what. If everyone went around taking everyone else's food, for example, mass confusion would result, if not outright violence! The same is true for housing. You can't have people arguing about the beds they will sleep in night after night. Now, simple, small societies can get by through informal rules or compromises, but this does not remain practical as societies grow. Therefore, formalized rules as to the owership of property have arisen. You see, ownership of something is no more than a formal way of saying that you have a right to use it. Each society must enforce these rules of property, and therefore there is some prohibition against breaking them, an activity we call "stealing.""

"OK, Dr. Newman, I understand *how* such a rule came about in our society, but why shouldn't an individual such as myself choose to steal anyway?"

"If you think about it, you wouldn't want another person to steal from *you*. If everyone did this, society would collapse. Since you don't want others to steal from you, you must encourage the prohibition against stealing in our society. Therefore, you should not steal yourself."

"Just a minute. I agree that I don't want others to steal from me, and therefore that I should encourage the prohibitions against theft in general, but what if I can find a way of personally stealing and not getting caught? I get the benefit of others not stealing from me, while I also get what I want!"

"Larry, I think that you must admit that this would be terribly unfair . . ."

"Of course it is unfair, professor, but so what? Why is it wrong to be unfair? I know that you will tell me society must encourage fairness for all sorts of reasons, but, again, why shouldn't I choose to be unfair if it benefits me and I can get away with it?"

"Larry, you are beginning to sound like a sociopath."

"OK by me. Call me what you will. What's *wrong* with being a sociopath?"

Before this gets too depressing, let's think about what has happened. Professor Newman has been facing a classical problem in ethical philosophy: can you derive what *ought* to be from what *is*? He is trying to explain that stealing is *wrong*, but he is succeeding only in explaining how such a prohibition may have come about in society. The good professor, working only from a description of what he can observe scientifically in the world, is totally unable to explain why the action is wrong.

Is it morally wrong to shoot a character in a video game? Of course not, for there is really nobody there. There is only a machine programmed to act in a particular way that we interpret as a bad guy. What about torturing a cat for sport? Science tells us about the organs that make up a cat. It explains about the nerve cells in its brain. It describes the reaction of the cat to various stimuli, which we call pain. From a scientific viewpoint, a cat is a complex machine that is programmed to react by screaming when you step on its tail. On the level of what *is* there is no essential difference between a cat and a video game. Perhaps torturing a cat is no more wrong than shooting the bad guys on the screen. After all, there are societies that condone the torturing of animals for sport; that is what a bull fight is all about.

Medical doctors are often asked about the morality of abortion. They are indeed qualified to tell us about the development of the fetus within the mother. They can tell us when the heart starts to beat, when the fingerprints are formed, and at what age the fetus appears to sleep and wake. But are they qualified to tell us whether abortion is moral in a given case? Don't we need to know something other than what *is* to know what *ought to be*? Science cannot teach us this; the answer must come from somewhere else.

We can identify with Ivan, the troubled character in Dostoyevsky's *The Brothers Karamazov*. Unable to derive *ought* from *is*, he concludes that if there is no God, then all is permitted. But we have a deep belief that stealing and causing pain are wrong. We *know* that all is not permitted. The conclusion is inescapable. There is some source of morality other than that which can be scientifically observed. There is some way of knowing about the world in addition to the scientific method.

4

The User's Guide
The Nature of Torah

Twenty years ago, a person buying a wristwatch received a nice box and perhaps a warrantee card, but no instruction manual. He did not need one. Everyone knew how to work a watch. He wound it once a day and pulled out the little knob when he wanted to set it. But today, things are different. It is not unusual for a digital watch to come with a thirty-page instruction manual, and there are still people who walk around with their precision timepieces off by seven minutes because they have forgotten which buttons to push.

It is not only wristwatches that come with user's guides. Microwave ovens, cameras, sometimes even toothbrushes now come with instructions. The user's guides are written by the manufacturers who designed and built the products, and they tell us how to get the most out of them, how to use them in the most beneficial manner. The universe, too, comes with a user's guide. It has also been written by its manufacturer. We call it the Torah.

The user of a wristwatch is not interested in how it works.

21

He just wants to know the time. Therefore, the user's guide doesn't describe the design of the electronic microchip within it, nor how the numbers are displayed on its face. It tells which buttons to push and in what order to push them. Similarly, the purpose of the Torah is to teach us how to lead our lives, not to teach us science.

That is not to say that the Torah doesn't recognize a deep need within us to explore and understand the physical world. Indeed, the oral tradition tells us that it was contemplation of the physical world that led Abraham to conclude that it had a creator. In the Book of Psalms[1] we see that King David viewed the physical world with wonder and a feeling of being overwhelmed by its beauty: "O Lord, our Master, how glorious is your name in all the world; whose majesty is set on the heavens . . . that I see Your name, the works of Your fingers, the moon and the stars that You have established." But the Torah realizes that we are capable of studying the physical world through the scientific method, on our own. It is not the purpose of the Torah to teach us what *is*.

If we did not realize this, we might pick up the written Torah with a different purpose in mind. Reading from the beginning, we would come upon a detailed description of the creation of the universe. We might conclude that we had before us some sort of science textbook dealing with cosmogony, the study of the origin of the cosmos. We would be wrong.

This is stressed by the eleventh century biblical commentator Rashi in his very first comment on the entire Torah. Rashi quotes his father as asking why the Torah begins with the story of creation. What relevance does the history of the universe's origins have to the purpose of Torah? Shouldn't it

---

[1]Psalms, chap. 8.

begin with the first commandment given to Israel, the first information concerning what is right and wrong? Rashi answers that the creation story has a legal purpose. It is evidence of God's manufacture of the Earth. Since God created the planet, he owns it and can give parts of it to whom He chooses. He gave it to the Children of Israel. This is the basis for the claim of the Jewish people on the land of Israel.

Aside from Rashi's almost prophetic anticipation (which he specifically states) that there would come a time when the nations of the world would collectively accuse the Jewish people of stealing the land, there is another remarkable point in his commentary. It is clear that he regards the principal purpose of Torah as moral instruction. Why then do we find history, personal biographies, or even descriptions of nature within it? Such narratives are included only when they can teach us a lesson about how we should behave.

For example, in the book of Genesis we find the story of Joseph being sold into slavery by his brothers. While sitting down to eat, the brothers see a passing caravan and decide to sell Joseph to the Ishmaelites who are leading the camels. These camel drivers then take Joseph into Egypt. Simple enough, but in telling this story, the Torah brings a seemingly irrelevant detail. It tells us that the caravan was carrying sweet smelling spices: "balm and labdanum." [2] Why would the Torah, in which every word is significant, tell us this detail? What could it possibly teach us?

Rashi gives us an explanation. He tells us that most caravans carried tar and pitch. This was an unusual one that did not carry these foul-smelling substances. Thus, Joseph was not forced to endure unpleasant odors as he was brought into

---

[2]Genesis 37:25.

Egypt. Yes, God wanted him to be brought to Egypt, but God did not force him to endure suffering beyond that which was absolutely necessary to the divine plan.

This has relevance to our lives. It is a basic principle of Judaism that we are to emulate God's qualities as far as we are humanly able.[3] God does not cause unnecessary unpleasantness, even when he must be firm. Neither should we. Thus, for example, when we discipline our children, we should do so only to the extent absolutely necessary, not any more than this. Yes, there are times at which we must express displeasure or even anger, but we should always be acting out of love, always acting with precision, and certainly never out of control.

We could go on. There are a myriad of lessons that could be learned from this story and from the particular words that are used to tell it. We are not being entertained with an interesting drama; we are being taught right and wrong. That is the nature of Torah.

The *Zohar* (the primary written text dealing with the esoteric parts of the Torah) quotes Rabbi Shimon as saying: "Woe to the person who claims that the Torah's purpose is to present wordly narratives and everyday affairs. Were that the case . . . we could construct a teaching . . . more praiseworthy . . . . If [the Torah's intent] is to present wordly matters, even temporal rulers have recorded more exited matters . . . . But, rather, the words of the Torah are higher matters and higher secrets."[4]

We might add: "Alas for the person who thinks the Torah is a physics textbook." The Torah has no real interest in natural history as a discipline in its own right. That is left to scien-

---

[3]See the examples given in *Mesechta Derech Eretz*, chap. 5.
[4]*Zohar Behaloscha* 152a, translated by Yaacov Douid Shulman.

tists. The Torah is not really interested in what *is*, its purpose is to tell us what *ought to be*.

As with all analogies, the comparison of the Torah with a wristwatch user's guide is not perfect. If we are given a wristwatch we could, in principle, take it apart to see how it works. From what the wristwatch *is*, we could eventually figure out how to use it. We have seen that this is not the case with moral obligation. We cannot learn an *ought* from an *is*.

Let us see this another way. We will consider a law found in the Mishnah,[5] the core text of the Talmud. We have an obligation in Torah of *matanos aniim*, gifts to the poor. When a farmer harvests his field, there are certain contributions he must make to those who are without food. Among these is *ma'aser ani*, the tithe for the poor. During the third and sixth years of the seven-year *shmitta* (sabbatical year) cycle, the master of an agricultural field must give one tenth of its produce to poor people. The *Mishnah* considers the case of two destitute workers who contract with a farm owner to harvest his crops. Each agrees to harvest one field in exchange for a fixed fraction of the produce. During this harvesting, the poor men function as masters of the fields and are therefore required to give a tenth of their crops to someone who is poor. Yet they themselves are poor, and it would be a hardship for them to give up this food. What can they do? The *Mishnah* suggests that they give their tithes to each other!

Just a minute! Isn't this cheating? No, the Torah tells us that it is not. It is actually an observance of Torah law. Each man has performed a *mitzvah* (an act desired by God) and the performance of a *mitzvah* has major effects. Rabbi Shlomo

---

[5] *Mishnah Peah*, chap. 5, number 5.

Twersky (of blessed memory) has pointed out[6] that the performance of a single *mitzvah* by a Jew with truly selfless intention can have untold consequences in the world. We have here a case of two men who have performed *mitzvahs*, the world has benefited, and they have each accumulated individual merit. Yet after the performance of these *mitzvahs*, equal amounts of produce have been exchanged between them and the world is in essentially the same state it was before the transaction took place.[7]

The Torah indicates that an event has taken place that has improved the universe, but a scientist observing the physical world would say that nothing significant has changed. Clearly, we cannot learn what is morally valuable only by understanding the physical world.

So far we have been concentrating on a single aspect of Torah, Torah as information transmitted by God to man. This is the aspect of Torah that we encounter in our lives. But the oral tradition indicates there is something deeper. The Talmud tells us that God created the Torah before the physical universe and that He looked into the Torah to know how to

---

[6]Rabbi Menachem Goldberger, personal communication.

[7]I am writing these words on the forty-second day of the *omer*, 5750. No one who observes *s'firas ha'omer* and encounters the progression through the *sefiros* from day to day can doubt that there are real effects taking place within the higher levels of reality. So, in a sense, the performance of *mitzvahs* does have a deep effect on what *is*, and a truly sensitive and learned person might be able to derive what is right or wrong from an understanding of these hidden levels. But the hidden levels are not observable in a scientific sense. For those of us who are not qualified to talk about *nistaros* (hidden matters), it is only the observable world that we can deal with and, in this book, when we speak of that which *is* it is the *olam niglah*, the readily observable world, that we are talking about.

create the world.[8] We have to be careful here. We are dealing with anthropomorphic language to attempt to grasp something that is truly beyond our understanding. Certainly before the universe was created there was not a scroll floating out in space somewhere. The closest we can come to describing what happened is to say that the ideas of Torah, the *mitzvahs*, first came into existence and that the world was then created so that they could be put into effect. The Torah is the blueprint of creation.

This is a radical concept. To understand the implications of it let us consider the *mitzvah* of honoring out parents. Why is it right to honor them? The social scientist might suggest a reason, that respect for elders contributes to the stability of society. But we now recognize that while this gives a reason why this rule is found in so many societies it does not explain why it is *right*. A moral philosopher may explain that we should honor our parents because they brought us into the world and cared for us, and it is moral to recognize when good is done for us. This is closer to the truth, but we have really only explained an *ought* by appealing to another moral value, which is really the best we can do. But if the Torah is the blueprint of creation, we are really asking the wrong question. We should be asking: "Why do we have parents?"

God is omnipotent. He could have created the world in any manner He wished. It could have had a fixed number of people who never died. Or people could have grown spontaneously from the ground. But in choosing how to make the world, the *mitzvah* came first. Before the world existed, God created the Torah and within it is the *mitzvah* of honoring our parents. The world was created so that the Torah could

---

[8]*Mishnah Avos* 6:10; *Bereshis Rabba* 1.

be put into effect. Therefore, God made parents so that we may honor them! This is indeed a radical idea.

When we study the biblical story of creation in greater detail, we will see that the sun and moon were created on the fourth day. The Torah gives a reason for their creation. They were not brought into existence to provide light and heat for the earth or to provide inspiration for poets. The Torah specifically states that they were created so that man could observe them and know how to count the days and years. That is, they enable us to establish a calender so that we may know when it is *Shabbos* (the Sabbath) when it is *Yom Kippur* (The Day of Atonement). The sun and moon were created so that these *mitzvahs* could be put into effect.

We have already learned that we cannot produce an *ought* from an *is*. The tradition does seem to hint, however, that God created all that *is* from what *ought to be*.

We have made some progress. Our path has been complex, but it is apparent that science and Torah deal with the world in very different ways. Their methods differ. Their purposes differ.

What is the purpose of science? Its goal is to understand the physical world. There are two motivations for this endeavor. The first is practical: by understanding the world we can manipulate it and use it to improve our lot. But there is a more overwhelming drive. Most scientists would admit to an intense curiosity about the world. They want to know how it is put together, how it works. They want to poke it, to play with it, to manipulate it. Whether this knowledge is useful or not is of no concern to them.

The purpose of Torah is to teach us how to live our lives—what we are to do in the world. It is to provide the information that we cannot derive from observation of the physical world. If science tells us the *what* of the world, Torah tells us the *why*.

Because of their difference in purpose, we are not really surprised that science and Torah express their ideas in different ways. The situation is much like the proverbial blind men examining an elephant: one touches a leg and perceives it as a tree, another encounters the tail and thinks it is a rope. Because of their differing orientations, they come to apparently different conclusions, but our deeper knowledge explains why each is correct in his own way.

Science and Torah are both valid ways of knowing the world. We can understand that sometimes they seem to say different things. But what about when they seem to be in direct conflict? What then?

5

A Concept from the Talmud
Reacting to Problems

The Talmud is an enigmatic work. To many, it doesn't seem to be a religious text. Where is the dogma? Where are the basic truths? Where does it tell you what to believe? And, finally, doesn't anyone ever agree with anybody else?

Rabbi Mottie Berger[1] has observed that there are two major criticisms voiced about the Talmud. First of all, the Rabbis are always arguing with each other. Second, they are always arguing about the most minute details! How can any of this be taken seriously? But taken together, these two criticisms answer each other. The Rabbis argue only about the details, not about the basic tenets of the Torah. There is never any disagreement about whether pork is kosher or whether stealing is permitted. Through the revealed Torah, God has given us clear communication about all these matters. Little time is spent discussing them in the Talmud. It just isn't necessary.

---

[1]Lecture given at Aish HaTorah's Discovery Seminar.

What the Talmud does concentrate on is that which is unclear, that which is problematic, that which *needs* discussion. But if God gave us the Torah, why are there ever any questions about what it contains?

One reason is that humans are imperfect. They forget things. Even Moses forgot a number of the basic Torah laws that God had given him.[2] He solved this problem by seeking guidance from God (to whom he could speak face to face), but in later generations scholars could not do this. Information was lost. Errors occurred in the transmission of the oral tradition. This was unfortunate, but unavoidable.[3] Indeed, it was recognition of man's fallibility, and fear that the oral tradition would be lost by future generations, that led Rabbi Yehudah HaNasi to compile the *Mishnah*, the central core of the Talmud that summarizes and records the orally transmitted material.

But even if information had not been lost there would still be uncertainties and disagreements. This is because the Torah was not presented to us as a well-defined set of principles. Rather, it was presented as a collection of paradigms, examples from which we could learn concepts that could be applied to new cases.

To tell us what is forbidden on *Shabbos*, the Torah does not give an abstract set of rules. Instead, it gives us a list of

---

[2]They are listed in M. Weissman, *The Midrash Says, The Book of Vayikra* (New York: Benei Yaacov, 1982), p. 337.

[3]After all, people must have the capability of forgetting. How else could they accomplish the *mitzvah* of *shichcha*, the law that crops forgotten during the harvest must be left behind for the poor to gather? (This is another example of the world being structured so that the Torah could be put into effect, God's production of *is* from *ought*.)

thirty-nine prohibited activities. From these, we are to derive the rest.

For example, one of the thirty-nine is the activity of threshing. This is a process in the production of wheat in which the grain is beaten, pounded, and otherwise physically abused to cause the wheat to separate from the chaff. Most of us are not farmers. We don't do a lot of threshing. Does this example have any relevance for our lives?

It does. Let's think of a nice cup of tea. It would really taste good with a little lemon juice in it. May I squeeze a lemon into it on the Sabbath? This activity does not seem to be among the thirty-nine examples given in the oral Torah. Yet when we compare lemon squeezing to wheat threshing, we see that they have a lot in common. In both cases, force is used to physically separate something we want (wheat, juice) from something we do not want (chaff, lemon pulp). By comparing these activities, we see that indeed we may not squeeze lemons on *Shabbos* since any physical separation of food from something it is organically connected to is not allowed on *Shabbos*. We have extended the paradigm and learned an important principle. We can go further now and see that we may not milk cows on *Shabbos*, for the milking process consists of physically separating the milk from the udders of the cow. Quite a lot of information has come from the single paradigm of threshing.

Lawyers refer to sets of principles as *black letter law*. They call collections of paradigms *case book law*. Most agree that case study makes better lawyers. They learn to think rather than just to parrot statutes. God wants us to think about Torah. He wants us to talk about it, to ponder it, and to grow old with it. He wants us to apply it to new cases that have not been encountered before. So He gives us paradigms and minds and turns us loose.

Inevitably, we will run into difficulties and disagreements. But Torah scholars do not see these as negative. Analysis of disagreements is a valuable way of gaining new insights. Even when the Talmud is able to resolve a conflict, it meticulously records all viewpoints lest some future opportunity for learning and analysis be lost.

Conflicts are so common in the Talmud that there are technical Aramaic terms that are used to denote the various types. A *tanai*, for example, is a disagreement between two rabbis from the time of the *Mishnah* (roughly the first two centuries of the common era). While a *tanai* may make it difficult to determine what the correct law is, it does not cause us any logical problem. After all, it is possible for two rabbis to disagree. But there is another type of disagreement that is harder to deal with. It is called a *kashia*, which means, literally, a difficulty. A *kashia* is an apparent contradiction where one cannot logically exist. That is, both conflicting statements come from sources that must agree. Often, both statements are the opinion of a single person, and a person cannot disagree with himself.

Some examples may make this clearer. A man and a woman wish to marry. The Torah tells us that the betrothal begins when the man gives the woman an object that has monetary value, such as a ring, and she accepts it in front of two witnesses with the intention of marriage. Of course, there is also usually a whole lot of dancing and occasionally some really good hors d'oeuvres, but it is the valuable gift that really matters. How much must this object be worth? The Talmud records a disagreement between the scholars Hillel and Shammai. Hillel states that it must be worth at least a *peruta*, the smallest coin in circulation at the time (about the value of one cent). Shammai disagrees and maintains the object must be worth at least a *dinar*, a larger coin worth 192 *perutas*.

There is much we could learn by analyzing this disagreement. What concepts of monetary value are Hillel and Shammai arguing about? What implications are there for other contractual matters? Who is taking a stricter opinion? There is an interesting difference of viewpoint, but there is nothing enigmatic. Two scholars have reached different conclusions. It is a *tanai*.

Let us consider a more problematic contradiction. In the book of Psalms[4] we find the statement: "The earth is the Lord's and the fullness thereof." Yet, elsewhere in the book[5] it says: "The heavens are the Lord's, but the earth He has given to man." The first verse clearly states that the earth belongs to God, while the second says that it is given to man. Can they both be correct?

This book was written by King David, so there is a single source for the statements. We cannot explain the contradiction as a difference of opinion. We have an apparent contradiction between two statements by the same source. This is a *kashia*!

The Talmud discusses this *kashia* in tractate *Brachos*.[6] This portion of the Talmud discusses the blessings we are told to make before eating, drinking, or enjoying other pleasures of the world. When confronted with the above contradiction, the Talmud observes that the apparent contradiction can be resolved if we take into account the practice of pronouncing blessings before eating. When King David speaks of God owning the world, he is speaking a basic truth. All that is in it belongs to the Lord, and if we were to consume its pleasures without restriction, we would be guilty of theft. But God

---

[4]Psalms 24:1.
[5]Psalms 115:16.
[6]Babylonian Talmud *Brachos* 35a.

has given us a gift. He will allow us to enjoy the pleasures of His world freely if only we pronounce a blessing before partaking of them. When we say a blessing recognizing God's kingship over the world, ownership of the food or drink is transferred to us. The first verse we quoted speaks of the situation before a blessing is pronounced, the second of the situation afterward.

There are major implications of this. Not only does food belong to God, to be given to us only for proper use, but this is also true of the rest of the world. This includes our bodies; we cannot answer questions of euthanasia, suicide, and abortion by simply saying: "It's my body; I can do with it as I choose."

We have learned a lot by resolving this *kashia*. We have seen that the apparent contradiction existed only because of our incomplete understanding of the situation. Once we understood what was going on, the contradiction disappeared. In addition, by resolving the *kashia* we have come to a deeper understanding of the world. Far from being a threat to our beliefs, a *kashia* is an opportunity to gain new insights. Torah scholars learn to welcome *kashias* rather than dread them.

Kashias are not unique to the Talmud (although the Aramaic term certainly is). They also occur in the sciences, and here, too, they lead us to new discoveries. Let us consider an example in the history of physics.

From the very early days of man's encounter with the world, researchers have wondered about the nature of light. The great physicist Isaac Newton thought that light was composed of small material particles, much like tiny microscopic bullets. Another English physicist, Thomas Young, hypothesized that light was not composed of particles at all, but rather of waves that move through space much as waves travel across the surface of the ocean.

If this were just a matter of a disagreement between two intelligent gentlemen, we would not have a *kashia*, merely two differing opinions. But this is not the case. Generations of physicists have performed careful experiments to test both of these conflicting theories, and compelling evidence has been found confirming each of them!

In the early nineteenth century, Young noticed that waves behave in certain characteristic ways. They tend to spread out in space, rather than remaining in a local region. This is called "diffraction." In addition, waves coming together from different directions will "interfere" with each other. At certain places they will cancel each other out, while at others they will add together and cause a large effect. None of these behaviors can occur with particles. Young's experiments showed that light produces both diffraction and interference. Therefore, it must be a wave; it cannot be a particle.

By the beginning of the twentieth century, other experiments had been done that showed the opposite. In one of these experiments (the "photoelectric effect," described in Chapter 23), it was observed that light shining on the surface of a metal would be absorbed only in small indivisible amounts. These minute parcels of energy came to be called "quanta." This is exactly the way in which particles behave— either they are absorbed by something, or they are not (there is no half way)—but it is not true of waves. Light therefore definitely consisted of small particles; it could not consist of waves.

Two sets of completely reliable experiments had shown contradictory things: that light consists of waves, *not* particles, and that light consists of particles, *not* waves. *Now* there was a *kashia*.

How did physicists react to this set of affairs. Well, they were troubled—but by and large they did not sink into depression, give up physics, or (more significantly) choose to ignore one of the two sets of contradictory experiments. Rather, they were intrigued. Clearly there was something new to learn here. Therefore, they started to work hard at trying to resolve the *kashia*.

How was this done? During the 1920s some young (and therefore unspoiled) physicists named Schroedinger, Heisenberg, and Bohr realized that scientists' intuitive concept of the universe was wrong. Because of this mistaken concept, when they looked at the experimental results, they encountered apparent contradictions. This problem could be solved only by radically altering the picture of reality in such a way that the wave and particle attributes of light were no longer contradictory. These physicists succeeded in constructing a new theory of the universe, now called the "quantum theory." We will discuss aspects of this theory in Chapter 12, but for now let us just point out that it changed the world view of physics so radically that the wave-particle paradox disappeared. This seemingly unresolvable *kashia* in physics was resolved. The ultimate outcome of the *kashia* was a reframing of our picture of reality. This readjustment has become the basis of all modern physics. It would not have come about if physicists had shied away from tackling the paradox.

When we compare the Torah and scientific viewpoints of creation, we will find some striking parallels. But we will also find some apparent contradictions. Some of these we will not be able to resolve with our present state of knowledge. How then will we react?

We will welcome the contradictions as opportunities for future insights. We will recognize that, if Torah and science are both valid, ultimately contradictions cannot exist. But that will not mean that we have all the answers. In truth, we can only understand the shallowest layers of Torah, especially when trying to plumb the depths of creation. In addition, the scientific picture of the world is constantly changing as theories are suggested, confirmed, adopted, and then perhaps overthrown by new data. Where we see a conflict today, we might see harmony tomorrow. Then again, where there is now consistency, there may later be an impenetrable conundrum. Old contradictions may disappear, new ones may arise, but the excitement and fascination continue.

This then is the central point of this book: a *kashia* is not a threat, it is an opportunity. As each problem is encountered, we learn. As each crisis is resolved, we move forward a little bit. We may not be able to show how it all fits together, but we can develop an approach to thinking about creation. It is the process that we seek, not the conclusion. Someday, we will get where we are going. Meanwhile, we can savor the voyage.

6

Approaching the Text
How to Read the Torah

U p to now we have been speaking in generalities. Enough
of this. It is time we got down to work. Let us begin to
approach the biblical text.

If we open the Bible to the book of Ezekiel we find that
the prophet has written a dramatic account of the revelation
he received from God.[1] Because he writes about the appearance of a heavenly vehicle, this is known as the Account of
the Chariot. Even a cursory glance at the text reveals that this
is difficult material. Ezekiel is trying to put into words an experience that just cannot be explained verbally. The information he has received from God is so esoteric that it is spoken
of only obliquely, and we have no hope of understanding it
unless we can find a teacher who is privy to the oral explanation that Ezekiel communicated to his students.

On the other hand, if we open the Torah and start reading the beginning of the book of Genesis, we encounter a story

------

[1]Ezekiel, starting with 1:4.

so easy to follow that we can even go to our local shopping mall and find an illustrated version for children. We might be tempted to think this is easy material. We would be making a big mistake.

The Talmud tells us to be careful. It warns[2] that the true meaning of the Account of the Chariot is so hidden that it should only be taught to one person at a time. It then goes on to caution that the true explanation of the story of creation should only be taught to two at once. Thus, the creation story contains within it material almost as esoteric as the Account of the Chariot. It is the most complex material in the Five Books of Moses.

To begin, we find in Genesis not one, but two, stories of creation. The book begins with a detailed account of the first six days. These six days take us step by step through the creation of the world and culminate in the creation of the first man. The account ends with the seventh day, the first *Shabbos*, on which God "rested." This takes us up to Chapter 2, verse 3. But now, the Torah seems to start all over again. In a more abbreviated form, it tells the story of creation in a slightly different way. Does the Torah wish to imply that the universe was created twice?

Certainly not. In asking this question, we have been making a rather cavalier assumption, one that is so built into our way of thinking that we are not aware of our bias. We have been assuming that the order in which things are written in the Torah is the order in which they happened. But if we operate under this assumption, we will quickly get into trouble.

As an example of this, let us consider an episode in the book of Exodus. Moses is confronting Pharaoh and demand-

---

[2]Babylonian Talmud *Chagiga* 11b.

ing that he release the Jews from slavery. Pharaoh is not taking too kindly to this. In a fit of rage, Pharaoh sends Moses away: "Go from me . . . for on the day you see my face [again] you will surely die."[3] He tells Moses to get out and never come back. He will kill him if he ever returns.

The Torah then goes on to describe a complex series of revelations that Moses receives from God and the reactions of the Jewish and Egyptian people. Only after telling all this does the Torah say about Moses, "and he went out from Pharaoh in fierceness of wrath."[4]

What is going on here? First Pharaoh tells Moses to get out; he will kill him if he returns. Next the Torah tells us of a whole sequence of seemingly unrelated events. Finally, it says that Moses left Pharaoh's presence. Did Moses leave after Pharaoh's threat, interact with God and the Jewish people, and then return a second time? If so, Pharaoh would have killed him. If the events are being presented in the order in which they happened, it does not seem to make sense.[5]

In the Book of Numbers, Rashi gives us an important rule.[6] Quoting the Talmud[7] he observes, "there is no before or after in the Torah." In modern terms: the Torah is not necessarily written in chronological order. Instead, it is written in conceptual order. Chronological order is used most of the time, but is freely ignored when it is necessary to make a point.

---

[3]Exodus 10:28.
[4]Exodus 11:8.
[5]Rashi is bothered by this, and he solves the problem by assuming that the revelation from God to Moses occurred in the presence of Pharaoh. But other commentators, such as R. Shimshon Hirsch, maintain that the verses are not in chronological order.
[6]Rashi on Numbers 9:1.
[7]Babylonian Talmud *Pesachim* 6b.

While this might seen strange to us, it may not have been so to speakers of biblical Hebrew. There is a theory credited to the psychologist Benjamin Whorf[8] that the language a person speaks influences the way in which he thinks (this hypothesis is sometimes called *linguistic determinism*[9]). In the English language, and most modern European languages, we find a definite sensitivity to the passage of time. We have three major verb tenses: past, present, and future. Yes, we have variations of these such as present perfect and pluperfect (whatever that is), but the basic division of time into three epochs is maintained. We, too, as English speakers think of ourselves as positioned in the present, with the past over with and the future yet to come. But not all languages have this division of tenses. The Hopi language, for example, distinguishes actions based upon their validity rather than their position in time.[10]

In biblical Hebrew there are really just two tenses.[11] The perfect tense indicates action that is completed, and the im-

---

[8]B. L. Whorf, *Language Thought and Reality* (Cambridge, MA: MIT Press, 1964), p. 252.

[9]S. Glucksberg and J. H. Danks, *Experimental Psycholinguistics, An Introduction* (Hillsdale, NJ: Lawrence Erlbaum Associates, 1975), p. 177.

[10]Thus, the Hopi expression *wari* means (in English) both "he runs" and "he ran," which are statements of fact, while "*era wari*" means "he ran," not factually but as a statement from human memory. Now that the reader is sufficiently impressed by my knowledge of Native American languages, it is only fair to state that I actually know nothing about such dialects. I have merely taken this example from Whorf's book. If you buy it, you, too, can impress your friends and readers.

[11]M. Greenberg, *Introduction to Hebrew, Second Edition* (Englewood Cliffs, NJ: Prentice Hall, 1965), chap. 9.

perfect tense indicates action that is still going on. There is really not even a present tense. Instead of saying "I write" the speaker actually puts together two nouns: "I, one whom is writing (*Ani Kotav*)." In modern Hebrew, the perfect and imperfect are used as the past and future, but to our ancestors they had different roles.

It is hard for us to imagine how a native speaker of biblical Hebrew thought about time. Perhaps that is why we don't really understand the rather strange grammatical phenomenon in which the Hebrew letter *vav*, which usually just means "and," reverses the tense of a verb following it. To Abraham and Moses this must have seemed quite natural.

In any case, we see that the presence of two sequential creation stories in *Bereshis* does not signify two creations. It didn't happen twice. Then why is it told in this way? This demands an explanation.

A quick look at the two stories provides a hint at what is going on. There are different names used for God in each of them. Why does the Torah use these different names?

The oral tradition tells us why. The different names indicate different ways in which God interacts with the world. When we experience God as the source of *chessed*, of loving kindness, we call him *Hashem* (actually, the unpronounced four-letter name of God). When we experience God as the source of *g'vurah*, of strength and justice, we call Him *Elokim*. In the first creation story, the name used is *Elokim*; in the second it is *Hashem Elokim*.

So why the two stories? Just as we perceive God in different ways, so do we perceive ourselves in different ways. The story of the creation is ultimately the story of the man's origin, and each of the stories has a different viewpoint. Rabbi Joseph Baer Soloveichik points out that man has a dual nature.[12] On one hand, we are biological beings, just as are dogs or monkeys. We have organs, require sleep and nourishment, reproduce, and die. There are people who live their lives on this level. Rabbi Soloveichik calls a human at this level *natural man*. But we have a second side. In the morning prayers we say, "and the improvement of man over the animal is nothing, for all is vanity, except for the pure soul [you have given us]." We have a soul, a spiritual nature, that distinguishes us absolutely from the animals. We are all capable of spiritual growth, although some choose not to pursue this route. It is not easy. Such growth requires us to experience the inevitable conflicts that morally sensitive individuals do when they interact with the world. Rabbi Soloveichik calls the spiritually growing individual *confronted man*. It is only confronted man who is capable of reaching the ultimate third level of the truly spiritual being. Although Rabbi Soloveichik sees both natural and confronted man as intertwined somewhat throughout the creation stories, each of them uses a different name for God since it focuses mostly on one of these aspects.

In the first story, the name used is *Elokim*, God as the source of law, the creator of a natural, ordered universe. Rav Avraham Kook sees physical and moral law as a continuum,[13] so in this aspect God is the source of both. God, perceived this way, created natural man, the biological being in harmony with the physical world. This is easy to grasp.

A more difficult idea occurs in the second story, and a much more enigmatic name is used for God: *HaShem Elokim*,

---

[12]J. B. Soloveichik, "Confrontation," *Tradition* 6:2 (Spring-Summer 1964): 5–29.

[13]A. Metzger, *Rabbi Kook's Philosophy of Repentance* (New York: Yeshiva University Press, 1968).

a dual name indicating both justice and mercy. Any parent can tell us about the conflict between these qualities. Upstairs is the 6-year-old child who has been put to bed early because of what he has just done. Mom and Dad can hear him whimpering softly and are experiencing the heart-wrenching pain that only loving parents know. He needs to learn a lesson, the quality of justice calls for this. But he is so *little*! Surely we should show mercy and let him come downstairs.

What to do? What to do?

The morally sensitive individual is confronted by such choices daily. As he seeks to unravel the riddles of moral choice, he understands why the Talmud tells us that *HaShem Elokim*, the One who embodies the tension between justice and mercy, cries out, "*Oy*, for my children." [14]

And so, the Book of Genesis tells of the creations of both aspects of the human condition. The opening of the book tells us the origin of natural man; the second story that of confronted man. We will now go through each of these stories in some detail.

---

[14]Babylonian Talmud *Berachos* 3a.

7

Natural Man
The Creation of Biological Man

Everything in the Torah is significant, even the shapes of the letters. The first letter in the Hebrew text of Genesis is a *Bais*, which is shaped similarly to an opening parenthesis. This tells us that what lies above, what lies below, and especially what lies before the beginning is beyond our grasp. We have no information about what came before the creation, so although we can ask many questions, we have no hope of ever answering them. What comes after the creation, however, is fair game. This, we will find, is difficult enough.

> In the beginning *Elokim* created the heavens and the earth. And the earth was *Tohu VaVohu* and darkness was on the face of the depths and the spirit of *Elokim* hovered over the face of the waters. And *Elokim* said, "let there be light" and there was light. And *Elokim* saw that the light was good and *Elokim* separated between the light and the darkness. And *Elokim* called the light "day" and the darkness "night," and it was evening, and it was morning, one day.

These are the first five verses in Genesis. We could spend the rest of our lives delving into them and still not gain more than a superficial understanding. Our investigation will be more limited. We will try to understand enough of the story to compare it with the scientific version of the same events. To bring some structure to this task, we will be writing on a blackboard. Our goal is to examine the beginning of Genesis and to list on the left side of the board the principal events in the creation. Once we have done this, we will construct a similar list, on the right side, summarizing the scientific viewpoint.

Let's begin. The Torah starts with, "In the beginning, *Elokim* created . . . " The verb used for "create" is *bara*, which is an unusual word in Hebrew. It occurs principally in only three contexts in the entire Torah. The first use is here, in which it refers to the creation of the universe. The second context is on the fifth day, where it indicates the creation of the great creatures of the sea. Finally, it is used to refer to the creation of man on the sixth day.

According to the Ramban and Sforno, this verb indicates a creation of something from nothing, *Yesh MeAyin* in Hebrew (or *creation ex nihilo* in Latin, which always sounds more intellectual). There is an important point here. According to these commentators, there were only three acts of creation that truly brought forth something from nothing: the initial creation of the universe; the creation of the first animate life, the sea creatures; and the creation of man.[1] All other acts were the transformation of something already existing into something else. Such transformations may or may not have been consistent with the laws of physics or biology as we know

---

[1]Ramban and Sforno on Genesis 1:1.

them today, but the events described by *bara* certainly were not. Creation of something from nothing definitely violates all known laws of science. The Torah is thus telling us that the initial creation was a truly unique event totally beyond our understanding.

Following this moment of creation, the Torah talks about the heavens and the earth, which we might identify as the spiritual and physical realities. The initial state of the physical universe is described as being *Tohu VaVohu*, an enigmatic term. The Talmud[2] attempts to deal with these terms separately. It identifies *Tohu* as a green or yellow line surrounding the world (the horizon, perhaps[3]) and *Vohu* as stones in the depth of the waters. These meanings are derived from the book of Isaiah, where it refers to "the line of *Tohu* and the stones of *Vohu*." [4] But Isaiah was a prophet, and the words of a prophet are attempts at verbalizing that which cannot be put into words. They do not constitute an attempt at explicating the literal meaning of the text (which is called *pshat* in Hebrew). Therefore Rashi does not consider this explanation in his commentary on the Torah, and neither shall we.[5]

Rashi does explain the terms in combination as referring to an astonishing emptiness. Rabbi Aryeh Kaplan translates

---

[2]Chagigah 12a.

[3]M. Jastrow, *A Dictionary of the Targumim, The Talmud Babli and Yerushalmi, and the Midrashic Literature* (New York: The Judaica Press, 1971), p. 1323.

[4]Isaiah 34:11.

[5]Similarly, we will not attempt to make a detailed comparison with the Ramban's explication of these words. Suffice it to say that he sees the *Tohu VaVohu* as a formless and malleable substrate from which all other matter was formed, quite similar to what we will say in Chapter 14.

them as "formless and empty." [6] It is hard for us to conceive of a nothingness as absolute as *Tohu VaVohu*, since we always think of nothing as the absence of something. Here we are talking about an emptiness so pervasive and primitive that it stands on its own merit. Rabbi Samson Raphael Hirsch describes it as "astonishingly chaotic," [7] and here we face a problem. How can nothingness be chaotic? Nothingness is an orderly state. Our mothers frequently complained that our bedrooms were messy, but not because nothing was in them! It was only when our clothes and toys were lying in mixed heaps on the floor that they got upset about the chaos. Chaos requires the presence of something, yet *Tohu VaVohu* indicates the complete absence of anything. This seems to be a *kashia*. We will discuss it further when we consider the scientific viewpoint of creation in Chapter 14.

We are next told that the spirit of *Elokim* hovered over the face of the waters. The difficulties here are endless. What waters? Where did they come from? "Over the face of the waters?" What did "up" mean when the solid earth had not yet been created? To take Genesis literally, we must know what it literally means, and the literal meaning of Genesis is far from obvious.

Rashi is also troubled by the problem. He responds to it with a new translation of the first verse—and a very important principle. Instead of "in the beginning, *Elokim* created the heavens and the earth," Rashi translates the opening verse as "in the initial stages of *Elokim*'s creation of the heavens and the earth." This seemingly small difference has major impli-

---

[6]Aryeh Kaplan, *The Living Torah* (New York/Jerusalem: Maznaim Publishing Co., 1981).

[7]S. R. Hirsch, *The Pentateuch, Vol. I Genesis*, trans. Isaac Levy (London: L. Honig and Sons, 1963).

cations. Rashi specifically states, "the text does not intend to point out the order of the acts of creation." [8] When describing the events within one of the first six days, the Torah is not giving a sequential account.[9] Even though the waters are mentioned in the second verse, we cannot yet say when they were created; it could have been later.

We have gone from a sequential picture in which the first verse marked the true beginning of the universe to a somewhat hazier picture in which we are not sure of the order of the various events. If we consider the first blessing of the evening service, we find a still more radically different viewpoint. We speak here of a God who:

> alters periods, changes the seasons, and orders the stars
> in their heavenly constellations as He wills. He creates day
> and night, removing light before darkness and darkness
> before light. He causes day to pass and brings night, and
> separates between day and night.[10]

This is all written in the present tense.[11] Here we find a concept of the continual renewal of the creation. The universe is constantly being created and sustained by God. Maybe the concepts described in Genesis describe not only events of the

---

[8]*Rashi* on Genesis 1:1.

[9]We might expect that, since the days are numbered, the events of the second day took place after the events of the first, but in general, as we mentioned in Chapter 6, the Torah is written in conceptual order rather than chronological order.

[10]R. N. Scherman, *The Complete Artscroll Siddur* (Brooklyn, NY: Mesorah Publications, 1985).

[11]We are using the modern Hebrew interpretation of tense, since it is most understandable to our English trained minds.

distant past, but also the underlying structure of the continuous daily creation. Just as the initial state of the universe was empty and chaotic, so the *Tohu VaVohu* must underlie the continuous renewal of creation.

This has all been a bit confusing. To make it clearer we will summarize by writing three things on the blackboard:

1. There was a moment of creation in which the universe came into being. It was created from nothing. This creation cannot be described by the laws of physics.

2. The creation is being continually renewed by God at all times.

3. In its initial state, the universe was both empty and chaotic, a paradoxical combination. Since the creation is being renewed constantly, this *Tohu VaVohu* must somehow underlie the present state of reality.

Well, so much for the first two verses. That was kind of difficult and abstract. Let's move on to something a little more concrete.

The next thing that happened was that God made a statement, "let there be light," and light came into being.

Altogether, God uttered ten statements in creating the world.[12] What is God doing going around making utterances? Who is listening? Of course, the Torah is using anthropomorphic language. God doesn't have a mouth or speak with sounds, but still we have difficulty in understanding what is going on. Some sort of declaration or formal consideration by God's intelligence seems to have been involved, but why

---

[12]Mishnah *Pirkei Avos* 5:1.

or how we cannot say. We will again consider this problem when we consider the scientific viewpoint in Chapter 16.

The first "stuff" to come into existence after the *Tohu VaVohu* was light. Since the word *bara* is not used, this was not a creation of something from nothing. The light must somehow have been a transformation of the *Tohu VaVohu*. To a nineteenth century physicist, this would not have made sense.

The physicist's universe has two things in it: matter and radiation. Matter can exist in four states: solid, liquid, gas, and something called plasma. Plasma is actually the most common state of matter in the universe as a whole, although there is not much of it on the surface of the earth. It is similar to a gas except that the atoms have been stripped of some of their electrons, so that everything in the plasma is electrically charged. The most frequent contacts most of us have with plasma in everyday life are the gasses within neon lights and the flames of our *Shabbos* candles.[13] The four states of matter correspond to the four primal elements of the ancients: earth, water, air, and fire.

In addition to matter, there is radiation. This includes X-rays, radio waves, and light. Nowadays, radiation is always produced from matter, and a nineteenth century physicist would therefore have said that the Book of Genesis is dreadfully wrong. How could radiation have been created first? In Chapter 17 we will see what modern physicists have to say about this.

God did not leave the primeval light around for long. The Torah goes on to say that He saw that the light was good and he separated the light from the darkness. Rashi quotes

---

[13]Which are pretty dilute plasmas, but plasmas nevertheless.

the Talmud[14] as explaining that there is an allusion here to something very strange. It says that God separated out some of the primeval light and hid it away for a future time when it would be used by the righteous. This light will extend through all space, and with it it will be possible to see from one side of the universe to the other.

We now have three more things to write on the blackboard:

4. God used ten utterances in the creation of the world.

5. The first component of the physical world to be formed through an utterance was light.

6. Part of the primeval light was hidden away for use in future time. It will extend across the entire universe.

The first day ends with a verse that almost seems designed to remind us how hard it is to understand the simple literal meaning of the text, even though it initially appears simple enough for a child. There was night and there was day, evening and morning. It seems so ordinary—until we realize that the sun was not created yet, and wouldn't be until the fourth "day." Evening begins with sunset, morning with sunrise. A day is a cycle of the sun. What did these terms mean before the heavenly bodies were formed? We cannot be sure. Yet, the Torah does use the word "day," and this word has a definite meaning to us. So, we will write on the board:

7. The universe was created in six twenty-four hour periods.

---

[14]Babylonian Talmud *Chagiga* 12a.

Finally, the first day comes to an end. The Torah says there was "one day," not a "first day." There is only a "first" when there has been a "second," and at this point the second day has not come yet. The Torah always uses language in a very precise manner. Here, at last, is something we can understand.

Let's move on to the second day. Only one thing was created. It is called the *Rekiah*, which is often translated as "firmament." This doesn't help us at all. We are not sure what the *Rekiah* is, but it does separate the "waters above" from the "waters below." Does the Torah have in mind the empty space in which we live, that which lies between the blue oceans and the blue heavens?

One of the greatest scholars of the deepest parts of the Torah was a man named Shimon ben Zoma. He is usually just called ben Zoma since he was part of a group of four major scholars in the time of the *Mishnah*, all of whom were named Shimon. He was one of the few people privileged with access to the parts of the oral Torah dealing with the creation. In the Talmudic tractate *Chagiga*[15], ben Zoma reports that he was amazed to find that the *Rekiah* is only three-finger-widths wide. Obviously, ben Zoma is not talking about either the sky or the space in which we live. He is talking about matters far beyond our understanding. We know that ben Zoma was eventually driven mad by his encounter with the inner meaning of this material.[16] If ben Zoma had difficulty understanding the *Rekiah*, we are unlikely to do so here. In fact, the Ramban strongly cautions us against hoping to understand it.[17]

---

[15]Babylonian Talmud *Chagiga* 15a.
[16]Babylonian Talmud *Chagiga* 14b.
[17]Ramban on Genesis 1:6.

Nevertheless, we have been given a statement in the written Torah, and we therefore have a right to ask what it means. Perhaps we can gain some inkling of what is going on by observing that the Talmud associates the Torah with water.[18] The *Rekiah* effects a separation between the waters above and the waters below. Is *Bereshis* indicating a separation in the world that reflects the distinction between the hidden and revealed parts of Torah? Is it perhaps a deep division between that which is proximate to us and that which we cannot understand? What kind of separation could this be?

The kabbalists speak of a deep primal separation between the ultimate reality and the world as we know it.[19] This separation is the result of an event called the *tzimtzum*.[20] The ultimate nature of it cannot be understood by mortal man. Nevertheless, we can approach the concept approximately.

God is an absolute unity, without attributes and without structure.[21] We declare this twice daily when we recite the *Sh'ma*. Far from being too complex for us to understand, God is actually too simple.[22] We are only able to think using concepts, ideas, relationships. When we face comprehension of

---

[18]Babylonian Talmud *Taanis* 6a.

[19]See, for example, R. A. Kaplan, *Inner Space* (Jerusalem: Moznaim Publishing, 1990).

[20]The *Torah Temima* [Genesis 1:6, comment on the words "*malei nima*" from Babylonian Talmud *Chagiga* 15a], in discussing the *Rekiah*, contrasts man's sloppiness in defining appropriate boundaries with God's precision. The verb he uses for precise specification is *tzamtzeim*, which derives from the same root as *tzimtzum*. Perhaps the *Torah Temima* is hinting at the identification of the *Rekiah* with the *tzimtzum*.

[21]R. A. Kaplan, *Handbook of Jewish Thought* (Jerusalem: Moznaim Publishing, 1979), section 2:7.

[22]Ibid., section 2:12.

an absolute simplicity, we are stuck, much as we are when we try to understand absolute nothingness. Any statement we can make, any idea we can have, any relationship we can establish is already too complex. We bog down in our inability to leave the bounds of this world of complexity and multiplicity to gaze at the pure unity and simplicity of God.

This leaves us with a question: if God is a perfect simple unity, how could He give forth a seemingly imperfect world of complex structure? Of course, since God is God, He can do anything. He is not bound by the limitations of our logic. But we are, and we are driven to ask the question and attempt to answer it as best we can.

And so we come to the idea of the *tzimtzum*. In human language, God contracted or condensed His perfect being to allow room for the imperfect world to exist. It is almost as if God created, for us, the illusion of his absence. The Hebrew word for the universe is *HaOlam*, which comes from the root *'oLaM*, meaning to hide or conceal. Literally, the universe is "the hiding place." It is the place in which God conceals himself and says "come find Me."

It is an exciting pursuit. Some people are expert participants, while others don't even seem to know that the search is going on. Abraham was able to poke below the surface of the manifest world and see God hiding. Occasionally we hear of scientists who gain some inkling of a Creator through the breathtaking beauty of astronomy or molecular biology. Unfortunately, however, many scientists develop an immunity to awe. Even among those who do take part in the search, very few have Abraham's sensitivity or the depth of his intellect.

What of the rest of us? We are faced with a game that seems impossible, but we need not lose hope. We have been given a license to cheat! It seems that God gave us a map that we can use to find Him. It is called the Torah. We have a long trip—actually it takes our entire lives and the road doesn't go straight, but if we keep returning to the map we are guaranteed ultimate success.

Since the *tzimtzum* involves the concealment of God's unity in the imperfect multiplicity of the world, the work of creation consists, to some extent, of the creation and maintenance of diversity. Perhaps this is the reason for the various *mitzvahs* that we have establishing separations of unlike things: milk and meat, wool and linen, and the entire area of *kilayim*, the avoidance of planting different crops together. Rather than the phrase from the American wedding service, "those whom God has put together, let no man split asunder," the Torah tells us, "those which God has caused to be separate, let no Jew join together."

What is clear is that, although God is the only force operating in the world, He conceals His presence and presents us with a multiplicity of less real forces. It is these forces that seem most real to us, since we are bound to the physical world, and it is only after we have broken the connection to this world and enter the afterlife, the World of Truth, that we can achieve a clearer picture of what is going on. It is time to write something new on our blackboard:

> 8. God concealed Himself in the universe so that, although the divine unity is the only true force, we imperfect creatures perceive a multiplicity of forces that actually seem more real to us.

This brings us to the third day. At first, the events of this day seem straightforward. God caused the waters below to be gathered together, and dry land appeared. He then commanded the earth to bring forth vegetation. This is one of the days that is easy to draw in children's books.

Regarding the waters, Rashi is bothered by two things.

The first is why the waters had to be gathered together in the first place. The second is that the Torah then calls the gathered waters seas, plural rather than singular. Rashi asks, "Isn,t the earth actually covered with one big ocean?" He answers the first question by explaining that the waters were gathered together because they were initially spread over all the earth. That is, the dry land was beneath the surface of the oceans. He answers the second question by explaining that the plural word "seas" is used because we humans think of the ocean as being broken up into different bodies of water according to local environmental factors. Rashi gives an example of this, that the fish in one location taste different from the fish in other locations. We all speak of the Atlantic and Pacific as being different bodies of water even though we know them to be connected. Simple enough. Rashi seems to be talking about the planet we are all so familiar with.

But there is one problem lurking here that makes it all come unstuck. If the dry land that appeared was just the familiar solid earth on which we live, where did it come from in the first place? Rashi seems to be saying that the only part God had in this was the uncovering of the land, not the creation or formation of it. But Rashi always speaks in simplified language, as does the Torah (*Torah b'Lashon Bnai Adam*, the Torah is written in the language of man). It has to be this way. How else could we teach it to our small children? Yet we recall that the true meanings are much deeper. What about the waters and the land?

Water has several meanings in Torah. It has the simple literal meaning of that common fluid we drink and wash with. In addition, as we have seen above, it also symbolizes the Torah. R. Aryeh Kaplan[23] points out that water also represents

---

[23]Unpublished notes.

the fluid or uncondensed state of matter. In the Psalms we see that the Hebrew word *mayim* can also refer to a vapor: "Who lays beams in the *mayim*, and makes clouds chariots." [24] What could be the meaning of the gathering of the waters? The verb root used here to indicate "gathering" is *KVH*. R. Samson Raphael Hirsch tells us that this refers to a gathering together caused by a striving toward a central point. [25] What was produced as a result of this condensation? One product was the appearance of multiple "seas," large bodies or clouds of fluid or vapor. The other product is what the Torah calls *yabasha*, which we have translated above as solid land. This word has a different literal meaning. R. Hirsch analyzes its root *YBSh* and finds that it is the diametric opposite of *mayim*, water. It indicates something dry and rigid. [26] Perhaps the best translation is "solid stuff."

Time, again, to write on the blackboard:

9. The fluid or vapor of the universe was gathered together.

10. This gathering was through a striving or force toward one place.

11. One product of the attraction was the formation of fluid bodies.

12. A second product was the formation of the first solid material.

---

[24] Psalms 104:3.
[25] S. R. Hirsch, *The Pentateuch, Vol. I Genesis*, trans. Isaac Levy (London: L. Honig and Sons, 1963), comment on Genesis 1:9.
[26] Ibid.

But still more happened on the third day. God commanded the earth to bring forth seeds, fruit trees, and vegetation;
and the earth did so. We might think that by the end of this
day, the earth was a beautiful garden, but we would be mistaken. Rashi, in commenting about the second creation story
(which we have not yet gotten to), follows the Talmud[27] in
telling us that plants did not actually come forth until the sixth
day.[28] At the end of the third day, what was in existence was
only the potential for vegetation: seeds or, in modern terms,
genetic material. So we finish the third day by writing:

13. God caused genetic material to come into existence.

On to the fourth day. Again, it seems simple. God caused
luminous bodies to appear in the heavens (the Torah says the
"*Rekiah* of the heavens"). He formed two principal ones, the
greater to rule by day and the lesser to rule by night. We identify these as the sun and the moon. He also created the stars
and other heavenly bodies.

Why are the sun and moon singled out? The Torah explains why God created the sun and moon. Their purpose was
not to provide energy for the earth, although the Torah does
state that the sun does so. After all, God could have created
an earth that did not need illumination. Rather, these bodies
were created so that human beings could look at them and
count the days and years, so that we could establish a calender
and know when it is *Shabbos* or *Yom Kippur*. Again, two principles are manifest in Torah: the Torah is written from the
human viewpoint; and the universe exists so that the Torah
can be put into effect.

---

[27]Babylonian Talmud *Chullin* 60b.
[28]Rashi on Genesis 2:5.

But were the sun and moon *created* on the fourth day?
The Torah does not use this term. R. Aryeh Kaplan comments
that there are three terms used in Genesis for creative acts.[29]
The first, the verb *bara*, we have encountered above. It refers
to creation of something truly new from nothing. The second verb, *yatzar*, means to form from something already in
existence. But it is the third verb, *oseh*, that is used in reference to the heavenly bodies. This indicates the completion of
an act. The heavenly bodies were not created on the fourth
day. Rather, they were completed on this day. Their place in
the universe was established. Thus, we write:

14. God completed the formation of the heavenly bod-
    ies.

15. The sun and moon were completed and positioned
    so that they could be used by man to establish a cal-
    ender.

On the fifth day, we find the first animal life. Let us examine the first two verses:

> And *Elokim* said, let the waters swarm with swarming
> creatures, living beings, and let birds fly over the earth,
> upon the face of the heavens. And *Elokim* created (*bara*)
> the great sea creatures, and all living creatures that creep,
> that the waters swarm with, after their kind, and every
> bird after its kind, and *Elokim* saw that it was good.

The first verse declares God's initiation, through an utterance, of the creation of the swarming creatures, the fish,

---

[29]Unpublished notes.

and the birds. But it is the second verse that is the more important one, since it uses the verb *bara*, the creation of something from nothing. The creation of the first animal life is the appearance of something truly new, something that did not exist before in the universe. The Torah repeatedly uses the phrase *nefesh chaya*, a living being, to describe these creations. The new addition to the universe is the *nefesh*, the animal life force or soul. We add to the blackboard:

16. God created the swarming creatures, the fish, and the birds.

17. The animal soul, a truly new addition to the world, was created from nothing.

Finally, we reach the sixth day. It starts quietly enough with the creation of the land animals. The verb *bara* is not used; the animal soul was already in existence, and land animals were formed from that which already existed. But a very dramatic moment follows:

And *Elokim* said "let us make man, in our image, after our likeness, and they shall rule over the fish of the sea, the birds of the heavens, the animals, all the earth, and over all the creeping things on the earth." And *Elokim* created (*bara*) the man in His image, in the image of *Elokim* he created him, male and female he created them.

Wow! The difficulties seem endless. Let's consider just two, the easier one first.

The second verse above cavalierly mixes singular and plural: "in the image of *Elokim* he created him, male and female he created them." How many people were created? To answer this, we must first contrast Hebrew and English vocabulary.

In English, the word *man* has two meanings. It means a human being (as in "that's one small step for man . . .") and it also indicates a male human being, as opposed to a woman. This causes English speakers all sorts of problems. No political party nowadays would consent to being run by a chairman. The leader must be a chairperson; and the party's literature must be delivered by letter carriers rather than mailmen. Hebrew doesn't have this ambiguity. The word *adam* indicates a human being, whether male or female, and the word *ish* indicates a male *adam*. On the sixth day, God created *adam*, and the written Torah does not explicitly state whether this being was male or female.

But the oral tradition does tell us. In the *Midrash*[30] we find that *Adam* was not either; he/she was both! In modern terms we call such a human being a hermaphrodite, a person who is both male and female. Hermaphrodites are still found today. They are quite rare. More often an apparent hermaphrodite is born that is actually male or female but has some abnormality that gives it the appearance of both, but true hermaphrodites do still exist. The appearance of one is considered to be a birth defect. *Adam* was both male and female. We can now understand the words: " . . . He created him, male and female He created them."

But, there is a far more troublesome problem in the verses above. God said, "Let *Us* make man." God is a total unity. Who was he talking to? We find what appear, at first, to be two opinions. Rashi and Sforno hold that God is talking to the angels; the Ramban explains that He is talking to the forces of nature. These apparently differing positions become con-

---

[30]*Bereshis Rabba* 8:1, quoted in R. Moshe Weissman, *The Midrash Says* (New York: Bnai Yaacov, 1980).

sistent when we consider the words of the Rambam: "for all forces are angels." [31]

In modern terms, God enlisted the forces of nature in His formation of the first man. This means that man was created in a manner consistent with the laws of physics and biology. If a scientist were to observe the process, he would not find anything out of the ordinary. Indeed, there is a source that tells us it took nine months for part of the first *adam* to be formed.[32] Why would God act in this way? It is one of the ways in which God hides in the universe. He must remain hidden so that we can have free will.

Many people hate to go to the dentist, but for others it is the dental hygienist who is the real terror. As she cleans their teeth, she lays upon them levels of guilt that even the stereotyped Jewish mother didn't dream of. "Do you want to keep those teeth? You must brush four times each day, and floss, and use dental stimulators, and water picks, and disclosing tablets, and plaque removers, and . . ." Now it's certainly nice for a person to have a hobby, but most of us don't really want to spend forty-five minutes with our mouths each morning.

Suppose that if a person didn't floss just once, all of his or her teeth would fall out immediately. How many of us would

---

[31]*Moreh Nevukhim*, quoted by R. Raphael Pelkocvitz, "Explanatory Notes," *Sforno* (New York: Mesorah Publications, 1987), *Bereshis* 1:26.

[32]This seems to be the plain meaning of the Jerusalem Talmud, which speaks of *Adam* as having components that took seven and nine months to form [Jerusalem Talmud *Yevamos*, 4:2]. It should be mentioned that the *Pri Megadim* takes this statement in the Jerusalem Talmud as referring to the future nature of man, rather than that of *Adam HaRishon*. But, if we find no problem with the plain meaning of the Jerusalem Talmud, there does not seem to be a need to include the resolution offered by the *Pri Megadim*.

floss?[33] Suppose that smoking just one cigarette caused a person to immediately collapse from lung cancer. How successful would cigarette advertising be? It is only because the effects of these indiscretions are so far in the future, so hidden from our immediate perception, that we are free to choose whether or not to commit them.

In the same way, if the presence of God were obvious to us, if we could clearly see the consequences of our actions, we would not have the free will to choose between good and evil. We would not then be able to grow, for it is only through the possibility of error that growth can occur. God hides Himself to grant us absolute free will, and it is this free will that is spoken of when the Torah says we were created in the image of God.

Thus God created man in a manner consistent with the laws of nature. In this way God hid, allowing a biologist to be an atheist if he wishes. There was no sudden puff of smoke followed by the appearance of a totally new yet somewhat poorly dressed biological being. Still, the verb *bara* is used to describe the origin of man. This does imply a sudden creation of something from nothing. How can we resolve this *kashia*? An additional problem lies in the general assumption of the oral tradition that the first *adam* was created as an adult. It is certainly not consistent with the laws of physics or biology for a person to begin his or her life as an adult. We have uncovered some internal difficulties in the text.

To begin finding an answer, we will consider a phrase from the Psalms[34] that calls the Torah "The word He commanded to a thousand generations." If we count the genera-

---

[33]I am grateful to Rabbi Dovid Gottlieb for suggesting this example.
[34]Psalms 105:8.

tions from Adam to Moses, we find that the Torah was given in the twenty-sixth generation of the human race. Why, then, does the Psalm speak of a thousand generations? The Talmud in Tractate *Chagiga*[35] answers that there are 974 missing generations before the first *adam* that might have been created, but were not. Although there are several ways of understanding these words, R. Aryeh Kaplan quotes traditional sources that indicate these 974 generations were human beings who existed before Adam.[36] We may identify these as biologically complete *homo sapiens* that were not implanted with human souls.

An important comment by Sforno takes us further.[37] He is reacting to the sequence of the first three verses describing the sixth day:

> And *Elokim* said: "Let the earth bring forth living creatures after their kind, cattle, creeping things, and beasts of the earth," and it was so. And *Elokim* made (*'oseh*) the beasts of the earth after their kind and cattle after their kind, and all that creep on the earth after their kind, and *Elokim* saw that it was good. And *Elokim* said: "Let Us make man . . . "

In the first verse, God utters his intentions to populate the earth with living creatures. In the second, the verb *'oseh* is used, indicating that the process was completed on the sixth

---

[35] Babylonian Talmud *Chagiga* 13b.
[36] R. Aryeh Kaplan, *Immortality Resurrection and the Age of the Universe: A Kabbalistic View* (Hoboken, NJ: Ktav Publishing 1993), pp. 7–11.
[37] Sforno on *Bereshis* 1:26 (I am indebted to Mr. Harold Gans for this observation).

day. Then comes the third verse. Sforno tells us that the creature *adam* is a species of "living being," one of those creatures created in the two preceding verses. When God said "Let
us make man . . . " and brought forth the totally new being
(described by the verb *bara*), He took an existing biological
being and placed the first human soul within it.

We have arrived at the following sequence of events
for the sixth day. Initially, God caused living creatures to
inhabit the planet. The details of this process, the sequence
in which the various species came into existence, is not known
to us in any detail from the text, but eventually, in a manner
totally consistent with the laws of physics and biology, a
species that we call *homo sapiens* came into existence. For 974
generations, this group of beings lived and multiplied in a
normal biological manner. How long was this period? If
we assume a generation lasts about twenty years, this would
have taken approximately 20,000 years. We cannot, however,
be sure of this. We know from the text of Genesis that the
first *adam* eventually lived for 930 years, and that the life
spans of the first generations of man routinely approached
large figures. If we were to assume a couple of hundred
years per generation, the 974 generations would have taken
about two hundred thousand years. In any case, God finally
chose one of these beings, a young adult who was also a
hermaphrodite, and installed the first human soul within him/
her.

Let's write on the blackboard:

18. God created the land animals.

19. Eventually, through a manner entirely consistent with
    the laws of biology, the species called *homo sapiens*
    appeared.

20. Initially, the life span of individual *homo sapiens* may have been quite long, often approaching one thousand years.

21. After a period of between 20,000 and 200,000 years, God selected a young adult *homo sapiens*, a hermaphrodite, and placed the first human soul into him/her.

We have now written 21 individual points on the blackboard. Without too much effort, we could come up with more. But we are getting tired. We really need a good rest. Thank God, one is coming: the seventh day, the one for which rest was created, the day that is the goal of creation. To understand this, we need a separate chapter, and a change of pace. Let's take a little time and relax. The next chapter takes us into *Shabbos*.

8

Shabbos
An Interlude

The seventh day was the first Shabbos. There was no physical creation; rather it was a day God refrained from creating. Therefore, it does not really concern us in this book, and many readers may be tempted to skip to the next chapter. But how could we discuss creation and not talk about Shabbos? A Jew simply cannot do this. So, we will take an intermission. We will introduce a guest writer, Mrs. Shana Goldfinger, who will give us a small hint of just what Shabbos is all about.

When guests come to my home to spend their first *Shabbos*, they invariably ask a myriad of questions about what they cannot do. It is rare that anyone asks what a person *does* do on *Shabbos*. Yet the essence of these twenty-four hours is a world that is created anew each week, one that renews each Jew's connection to God, to the Jewish people, to himself, and to the meaning of man's existence on earth.

To attempt to explain *Shabbos* in words is a lesson in futility. Some things can only be learned through experience.

However, I will attempt to do so in this chapter with the understanding that an intellectual analysis of *Shabbos* must be coupled with an experiential understanding to be complete.

Let's return to the beginning of the world. One of the plants that God created was the fruit tree. A seemingly simple, straight-forward production. However, the actual words used to describe this creation shed light on man's relationship to God at that time. The Torah states that God made "fruit trees which made fruit each according to its own kind." [1] In other words, an apple tree made apples, an orange tree made oranges, and so on. This does not seem very interesting until we look a little closer at the Torah's wording. We find that there is an extra word used. It states that God made "fruit trees" and then repeats the word "fruit" in the explanation that each tree specialized in one kind of produce. Why not say God made "trees which made fruit each according to its own kind"? Whenever an extra word is used in the Torah, it is a clue to stop and look for a further message beyond the obvious literal meaning of the text.

The commentary of Rashi is always the first step in the search for messages hinted at in unusual sentence structures or words. Rashi comments that the repetition of the word "fruit" indicates that "the taste of the tree was to be the same as the taste of the fruit." [2] In other words, if you bit into the trunk of an apple tree, it would taste like an apple. An interesting detail describing what ideal life would have been like in the Garden of Eden, but surely there were many other aspects of the world at the time that would have been just as important to record. Why choose this one? Perhaps because

---

[1]Genesis 1:11.
[2]Rashi on Genesis 1:11.

this phenomenon can be seen as a metaphor for what life would have been like for Adam and Eve in the Garden of Eden had all gone well.

What is the purpose of a fruit tree? A tree trunk grows and develops branches and leaves. It's true that trees provide beauty and shade and even perhaps wood to build a home—but when such a tree is planted it is not considered complete until it bears fruit. When the first fruits are harvested and are plentiful and tasty, the tree has completed the task for which it was planted. How did these fruits come about? Only through the efforts of the trunk, the branches, and the leaves. Without the rest of the tree, there would be no fruit. One could look at the body of the tree as the means for reaching a goal—the production of the fruit.

Originally, God intended the tree trunks to have the same taste as the fruit they would eventually produce. He intended the means and the goal to have the same taste—a unity of purpose within that creation. He also created a unity of purpose within man so he could understand his purpose in the world and discern the means of attaining it. Adam and Eve understood that their purpose was to be close to God and that the means of attaining that goal was to follow God's rules in the world. Their means had the same "taste" as their goal. The goal was clear, and so were the ways to reach it.

So, what happened? Apparently, God did not want a world of automatons mechanically doing His bidding. Man was to be a thinking being. To bring this about, God put free choice into the world. Free choice meant that man could choose to wander from the path that would ultimately bring him close to God. This concept of wandering from the correct path is the Jewish concept of *chait*, which is usually inadequately translated as "sin." The word "*chait*" literally means to miss the mark—much in the same way that one shoots an arrow and misses the target. When Adam and Eve wandered from the path by disobeying one of the rules of the world, the purpose of the world and the means for attaining it became blurred. Their unity of purpose and means was shattered, and they now had to go out into the world to seek it for themselves.

God, however, did not sent them out without a remedy. Just as God built into the world the possibility of *chait*, He also created the possibility of *t'shuvah*. The English translation usually given for *t'shuvah* is "repentance," but a more literal translation is "return." The remedy for *chait*, a straying from the path, is *t'shuvah*, a return to the path. When a Jew does *t'shuvah*, he takes stock of his goals, of whether the means he is using to attain these goals is working, and he maps out a plan to get back on the path that will take him where he wants to go. The punishment may be seen as the anguish a person feels when he realizes how far he has strayed and how much time and effort he has wasted on the detour.

Shabbos is a day when a person has a chance to stop, think, and re-evaluate the path he has followed during the past week to see how much closer it has taken him toward his goals.

The Talmud tells us that on *Shabbos* God gives each Jew an extra soul.[3] We welcome that extra soul with the lighting of the candles and usher it out with the *Havdalah* service. I never understood this concept of an extra soul until a wonderful teacher of mine, Rabbi Avraham Baharan, introduced me to Elijah the Prophet. Let me attempt to introduce you to him.

Elijah is a prophet who never died. Instead, he wanders the earth and is a visitor at every Passover seder, a guest at

---

[3]Babylonian Talmud *Betza* 16a.

every circumcision, and a rescuer of Jews in trouble. The two things everyone can always count on are taxes and death. Taxes were created by governments ostensibly to provide services for the people who pay them. Death entered the world as a result of Adam and Eve breaking that original unity of purpose in their relation with God. The path that God gave Adam and Eve to remain close to Him was to create bodies and souls that could use this world according to God's commandments. The bodies were the means for the souls' purpose of attaining closeness to God. When Adam and Eve did a physical act that was not in accordance with God's will, the unity of purpose that united their physical and spiritual parts was destroyed, and the body and soul were no longer able to act in unity. The natural result was the need for death. Death is an abandonment of the physical part of a person in this world so that the soul can reunite with God. It has become necessary for most of us since we no longer unite body and soul in a single clear purpose. I don't know if Elijah succeeded in avoiding taxes, but he was able to achieve a unity of body and soul and thus avoid death.

Elijah was the kind of person that many today might call a zealot. He had a vision and viewed people and life through this vision. Elijah could not understand how people could be noncommittal in their beliefs. Around him, the people of Israel were worshiping the idol Ba'al, which had been introduced by King Ahab's wife, Jezebel. But the people had not totally given up worshiping God as well. Elijah—a man who fashioned his life around his vision—could not understand how people could not make a full commitment to what they believed in no matter what the nature of that belief. Elijah thought that, if he could just prove definitively to the people the validity of God and His power, the people would commit themselves to a total belief in God and His Torah. Elijah asked God to give him one day to bring this proof to the people. He would challenge the priests of Ba'al to a final show-down on Mount Carmel. Then the people would see; then they would understand.

The challenge was simple. He and the prophets of Ba'al would build altars. Each would, in turn, call to their respective gods. The god who brought down fire from the heavens to consume the sacrifice on the altar would prove his existence and power to everyone.

The appointed day arrived. The priests of Ba'al built their altar, and Elijah built his. Elijah very graciously gave the priests of Ba'al the first chance. They prayed and shouted to Ba'al. They danced and chanted and performed whatever ceremonies they possessed to bring a response from their god. Elijah watched their efforts and taunted them. "Yell louder," he suggested, "maybe he's asleep!" Finally it became apparent that their efforts were in vain.

Next came Elijah's turn. To eliminate any doubts about his using trickery, Elijah dug a trench around his altar and filled it with water. He thoroughly doused the altar and the sacrifice with water to make things more difficult. Elijah then called to God to answer him just this one time and to show the people of Israel His power. A fire immediately descended from heaven and consumed the sacrifice, the altar, and all that surrounded it. Elijah was triumphant. He was sure that this contest had proven to the people God's existence and power. Surely now they would stop worshiping Ba'al and renew their commitment to God and Torah.

At first it looked as if Elijah were correct. The people shouted "*HaShem Hu Elokim—HaShem Hu Elokim*" (*HaShem* is God, *HaShem* is God), chased the priests of Ba'al to the river, and slaughtered them. It seemed as if Elijah had been victorious. Back at the palace, however, Jezebel was fuming.

She was the one who had brought the god Ba'al to Israel. Now look what Elijah had done! Jezebel sent a message to Elijah—the next day he would be killed.

But wait—she is the queen! Why not kill him right away? Apparently Jezebel had a better understanding of people than did Elijah. She knew that if she killed Elijah on that day, the people would turn against her. She would wait a day until their enthusiasm waned and then take her revenge. After all, it doesn't take long for people to fall back into their old patterns of thought and action.

Elijah understood immediately that her perception was true. He fled for his life to the desert. There, on Mount Sinai, where the Torah was first revealed, he hid and awaited God's instructions. A big wind came and furiously blew, but God was not in the wind. A big thundering came and filled the air with its noise, but God was not in the thunder. Finally, a small still voice was heard, and God's presence was felt by Elijah. God told Elijah that his days as a prophet were over. Elijah had become one of those very few individuals for whom the purpose in life and the path to attaining it are so clear and unconfused that they cannot imagine anyone making the mistake of turning off at one of the detours. His spiritual level had made him incapable of helping the people; he could no longer understand them.

Elijah appointed a new prophet, a new King of Syria, and, his work in this world being done, he walked down to the river to leave it. Elisha, the new prophet, followed him. What he saw was a fiery chariot that swooped up Elijah and carried him—body and soul—to God. Elijah did not need to leave his body on earth. His body and soul were united in purpose, as were the trees and fruits in the original concept of the Garden of Eden. He had elevated it to the level of a soul. In a sense, he made his body into an extra soul.

On *Shabbos* we, too, make out bodies into extra souls. During the week, we eat for sustenance. On *Shabbos*, the physical act of eating becomes a *mitzvah*, as it increases the enjoyment of the day. During the week, we dress for practical reasons. On *Shabbos*, the physical act of dressing becomes a *mitzvah* to lend honor to the day. On *Shabbos* we take the physical side of our existence and raise it to the level of the spiritual. This, then, is the meaning of the extra soul the Jew is given on *Shabbos:* it is the body, elevated and purified. On *Shabbos* we return to the ideal Garden of Eden where the taste of the tree was the same as the fruit, where the path to closeness with God was clear, and where man's spiritual and physical aspects united together to work toward closeness to God.

We exit *Shabbos* changed, inspired, and ready for the work of the week. So, as we leave this discussion of *Shabbos*, let us get back to work.

9

Confronted Man
The Creation of Spiritual Man

In Chapter 7 we stepped through the first creation story, that of natural man. There is still room on our blackboard. Let's go on to the second story, that of confronted man. In this story, the creator is perceived as *HaShem Elokim*, the God of law and mercy, and the man He created is troubled by the agonies of moral choice.

We found the first creation story to be just that: a story. It was a sequential series of events, and we were able to list them in order. We will find that the second story differs in two ways. In the first place, it is so rich with moral and spiritual ideas that we will be drawn into taking detours to discuss them. We will really have to work to stay on track. Secondly, we will find it much more difficult to discern the time sequence of the events.

Rashi reminds us that the second creation story does not come to teach the order in which the events took place.[1] In-

---

[1]Rashi on Genesis 2:8.

stead, the Torah is here written in conceptual rather than chronological order. What do we mean by this? What can we learn from the order in which the Torah tells the story?

To learn anything from the Torah, we make use of principles of reasoning transmitted to Moses as part of the oral tradition. There are definite rules that may be used to draw conclusions from the written text. As an example, there is a rule called *Klal uPrat*, meaning "a generality followed by a specification." This rule is used to derive *halachic* (legal) information from the written Torah. When the text makes a general statement and then follows it with one or more specific cases, the law being discussed applies only to the specific cases listed. That sounds pretty cryptic; an example will make it clearer.

We know that God has told us to bring animals as offerings to the Temple in Jerusalem (when it exists). What types of animals may we bring? The Torah says that offerings may be brought "from the domestic animals, from the cattle, and from the flocks (sheep or goats)." [2] Now, "domestic animals" is a generality. It includes all sorts of animals such as cattle, sheep, goats, horses, donkeys, and so on. But it is followed by specific examples: cattle, sheep, and goats. Applying the rule of *Klal uPrat*, we determine that Temple offerings are limited to the specific cases of cattle, sheep, and goat *only*.

If the order in the text is reversed so the general case follows the specific examples, we have a situation called *Prat uKlal*, a specification followed by a generalization. The Torah is then telling us to generalize the law to all cases. As an example of this, we consider the *mitzvah* of returning a lost object to its owner. The Torah says that we should "do thus

---

[2]Leviticus 1:2.

with his donkey, do thus with his garment, and do thus with all that your brother has lost." [3] Since the particular examples of "donkey" and "garment" are followed by the generality "all that your brother has lost," the *mitzvah* is generalized to apply to all lost objects.

Together, the rules of *Klal uPrat* and *Prat uKlal* demonstrate the principle we have spoken of above, that the order of words in the Torah is vital for understanding its conceptual content, but not necessarily relevant to the chronological sequence of events. But the above rules apply to legal material. Are there any similar rules that apply to narrative and homiletical material?

Rashi quotes a source from the time of the Talmud, R. Eliezer ben R. Yosi HaGlili, as listing thirty-two such principles.[4] One of them is called *Klal SheLeAcharav Ma'aseh*, "a generalization followed by a particular event." It is similar to the rule of *Klal uPrat* that we discussed above, except that it applies to narrative rather than legal material. This rule tells us that the particular event should not be viewed as having happened later in time than the generalization; rather, the Torah is bringing the specification to further elaborate details of the general event that is written about earlier in the text.

Rashi observes that the verses describing the creation of man in Genesis form such a sequence. The statement "and He created man" in the first creation story[5] is a general statement. The second story goes into greater detail, and therefore the

---

[3]Deuteronomy 22:3.
[4]Rashi on Genesis 2:8 with the gloss of Rev. M. Rosenbaum and Dr. A. M. Silbermann, *Pentateuch with Rashi's Commentary* (New York: Hebrew Publishing Co., no year given).
[5]Genesis 1:27.

two stories together form a general statement (that man was created) later followed by several particularly significant aspects of this creation. Applying the principle of *Klal SheLeAcharav Ma'aseh*, we see that the events of the second story did not happen at a later time, but rather that they are written later in the text to help us delve more deeply into the creation of man, mentioned so briefly in the first story.

These deeper aspects are crucial to understanding confronted man. They were not needed in the first creation story since natural man is a much simpler animal.

On to the details! While the first creation story could be taken to be a treatise on cosmogony (the study of the origins of the physical universe), with natural man a minor player in the drama, this is not possible with the second story. Confronted man is the central figure, and all of creation is seen to be but the arena within which he acts.

The story starts quietly enough with a brief summary to remind us what is going on: "these are the generations of the heavens and earth as they were created (*bara*) on the day *HaShem Elokim* made (*'oseh*) the earth and the heavens." [6] This sentence has immense scope. It covers everything from the initial creation of the universe from nothing (*bara*) through the completion (*'oseh*) of the physical and spiritual realms. It is a general, all-encompassing statement.

Following this synopsis, we are given a more detailed description of the events that follow. At first, the earth is barren. The potential for vegetable life lies within it, but plants have not yet appeared. Gently, God causes a mist to arise from the earth, and from this mist comes water, rain, which coaxes the hidden life to sprout. The stage is now set for man. He is

---

[6]Genesis 2:4.

formed from the dust of the earth, and for the first time we realize the relationship between the words *adam*, meaning man, and *adamah*, meaning earth. God breathes life into this creature; again we see the installation of a living soul.

The text tells us that man was "formed" rather than "created." But the spelling the Torah uses for the verb "to form" is irregular. This verb should be written *YTzR*, but instead it is spelled *YYTzR*. There is an extra Hebrew letter *yud*. Why is the word spelled with two of these letters?

In Hebrew, a 3-letter root can be used to construct a verb, a noun, or even an adjective. The language is very flexible. We are even permitted to make nouns from verbs and vice versa ourselves. Scholars of English would cringe if they heard us say "let's car to the restaurant and get some eats," but in Hebrew such constructions are entirely proper. Thus, although *YTzR* is a verb meaning "to form" it can also be viewed as a noun, in which case it indicates an urge or drive, that which forms our personality. The doubling of the *yuds* indicates that man was created with two conflicting drives, the *Yetzer HaTov*, the urge toward good, and the *Yetzer HaRah*, the drive that, if unguided, leads to evil.[7] We say "if unguided" because the Torah tells us how to turn this drive toward good. With the proper guidance, the same urge that might lead a person to become a murderer can instead, lead this person to become a surgeon.

Sigmund Freud's followers thought that he discovered this. He called it sublimation. In his view, we all contain within us a primitive and wild component of our unconscious personalities that he called the "id." In Freudian psychology,

---

[7]S. R. Hirsch, *The Pentateuch, Vol. I Genesis*, trans. Isaac Levy (London: L. Honig and Sons, 1963) commentary on Genesis 2:7.

mature, civilized people—meaning those living in Vienna—
sublimate or direct their ids into socially acceptable directions.
Thus the cultured and gentle family doctor is actually subli-
mating and hiding his violent dark impulses. We all know
what happened more than fifty years ago when a short, slightly
disheveled man with a small mustache found and released the
evil lying just below the cultured veneer of Austria, Germany,
and the rest of Europe.

But although Freud rediscovered the evil urge, his view
is far more pessimistic than that which the Torah has been
giving us for the last thirty-three hundred years. While Freud
maintained that it is only our external environment that forces
us to turn our ids toward good, the Torah tells us that in
addition to the *Yetzer HaRah*, we have a second even stronger urge, that holy spark within us that gravitates toward God.
This is the *Yetzer HaTov*, the urge toward good. It brings us
to Torah, the instruction manual that tells us how to use our
Yetzer HaRah for good rather than evil. The potential murderer becomes a surgeon, the potential thief becomes a fund-
raiser for an orphanage, and the potential seeker of attention
and self-aggrandizement becomes an author.

Rashi puts a different slant on the story of the two *yuds*.[8]
Viewing *YYTzR* as a verb, he explains that God engaged in
two acts of formation. Not only did He create our mortal
physical bodies, but He also created a part of us that is immortal. Rashi seems to be referring to an eternal part of our
physical selves, the bodies we will have in that future time
when the dead will be resurrected, but there is also a psychological side to this immortality. Rav Eliyahu Dessler teaches
that there are deep inner parts of our personalities that are

---

[8]Rashi on Genesis 2:7.

immortal, parts that remain unchanged after death.[9] In fact, Rav Dessler uses this information to help us understand the nature of the ultimate reward and punishment in the after-life. During life in this world we can grow and change, but this is no longer possible when we have left the physical world in which we have free will. After physical death, God is no longer so deeply hidden from us, and we cease to have the free will ability to change ourselves. Our personalities remain as they were in life. Thus, the person whose personality is focused on clothing or sensual pursuits is pained by the loss of a physical body and experiences the afterlife as a punish-ment, while the person emotionally invested in Torah is fi-nally able to achieve higher levels of understanding than were ever possible during life and so feels deeply rewarded.

From the verb *YYTzR*, we learn that man has mortal and eternal aspects, both spiritual and physical. From the noun *YYTzR*, we learn of the two drives within us—the drive to-ward good that focuses on the eternal and the drive that can lead to evil if we permit it to focus on the mortal and tran-sient.

Wow! Look how much we have learned from a single little extra letter! When we take up the story of confronted man, we are quickly drawn away from a simple physical de-scription of what went on during the creation. We are inexo-rably drawn into deep questions about the meaning of it all. However, we must resist this temptation, for our topic is the scientific view of creation. Reluctantly, we leave our consid-eration of *ought* and return to the realm of *is*. We summarize what we have learned by writing on the blackboard:

---

[9]E. Dessler, *Strive For Truth* (New York: Feldheim, 1978), p. 73.

22. God caused a mist to arise from the earth. The mist produced rain that watered the earth.

23. God formed man from the dust of the earth.

While natural man is no more than a particularly advanced animal, confronted man is something very different. Thus, while natural man and animals are both described as *Nefesh Chaya*, living beings, the second creation story tells us that *HaShem Elokim* breathed into man a *Nishmas Chaim*, a soul unique to man. This soul is capable of cognition, conceptual thought, speech, and higher spiritual functions. It is this soul that is being referred to when the Torah tells us we were created in the image of God.

Before the appearance of this new capacity, all was harmonious and simple in the world. But immediately after the installation of the human soul in man, God presented him with his first problem. Things were not going to be so easy anymore. The entire saga of human history was about to begin.

The saga begins in a deceptively simple manner. After God caused plants to grow from the earth, he formed a beautiful garden and placed man within it. Man was then given a single *mitzvah*: he might eat from all trees in the garden except one. The fruit of this tree was forbidden to him. Man was given his first moral challenge.

God observed that, in his new circumstance, it was not good for man to be alone. Therefore, all the other animals, which God had also formed from the dust of the earth, were brought before him to see if he could find a mate.

There is a lot to analyze here. Let us begin with the order in which things are mentioned. The second creation story speaks first of man and then of the plants and animals—an order different from that in the first story. This does not disturb us.

We know that the Torah is written in conceptual, rather than chronological, order. Natural man is the most advanced species of animal, but little more. Therefore, in the first creation story, we are told of his formation as the last of a sequence of increasingly complex physical productions. Why, then, does the Torah tell us of confronted man's creation before discussing the plants and animals? The reason is that confronted man is a spiritual being and the rest of creation was created only to be an arena for his action, to serve him in his growth toward ultimate perfection. It is because of man that the animals exist; it is to feed man that vegetables exist. Therefore, confronted man's creation is spoken of first.

What of the plants? Biologically, they are more primitive than animals. Yet in the second creation story we are told of their creation before that of the beasts. It seems strange, but if we consider what is really going on, it all becomes clear. The first man was placed by God in the midst of a beautiful garden, an environment of vegetation. He was allowed to eat whatever he wanted, with one exception. The fruit of one tree was forbidden. The first man had a single *mitzvah*, and it was realized through vegetation. The vegetation presented man with a moral challenge; the animals did not. Since the performance of *mitzvahs* is the entire reason for man's existence, the creation of the plants, which enabled the first *mitzvah* to be realized in the world, is described first.

What of this first *mitzvah*? What was there in the nature of this tree that led God to forbid man to eat from it? According to R. Shnayer Lewis, absolutely nothing![10] In his analysis, the forbidden tree did not differ from any of the other trees in the garden. It did not differ in appearance, in aroma,

---

[10]R. Shnayer Lewis, private communication.

in taste, nor in its physical nature. The only thing different about this tree was that God said, "Don't eat from it." There was a *mitzvah* associated with it. When the first man and woman ate from the tree, they experienced for the first time their ability to disobey God. They came face to face with the potential for evil within themselves. This was the knowledge that came from eating the fruit of this tree, and the reason it was called the tree of *knowledge* of good and evil. It could just as well have been any other tree or object that God forbade. We have here a clear example of an *is* proceeding from an *ought*: first came the *mitzvah* not to disobey the will of God, and then the tree was created so that the *mitzvah* could be put into effect.

Oops! We are slipping into moral issues again. Not very good scientists, are we? We had better write again on the board:

24. God caused vegetation to sprout from the earth.

25. God formed the animals from the dust of the earth.

In the garden, confronted by his single challenge, man was lonely. God saw that it was not good for man to be alone. So far, so good. But what came next is a little strange. God brought all of the animals before man to see if he could find a mate, and he could not. The Torah says that *Adam* could not find an *ezer knegdo*, a helper *against* him.

There are two things we must understand. In the first place, what is meant by an *ezer knegdo*? We can understand a helper who is *with* us, but what about one *against* us? It all boils down to the difference between Hollywood and life.

The movies tell us that we can live happily ever after. We are told that in a perfect marriage two hearts beat as one, and you never have to say you're sorry. When it doesn't work out this way, we feel that something is wrong. Any family counselor will tell us of the basically OK couples who come in suffering from nothing more than a case of unrealistic expectations. What they usually achieve in counseling is they learn that no marriages are perfect, and they come to accept this and get on with their lives.

The Torah viewpoint differs from this in a very subtle way. Yes, real marriages are not without conflict, but this does not make them imperfect. Indeed, the *perfect* marriage does contain disagreement. Just as argument is seen, by the Talmud, as the way to eventually uncover what is true, so is honest dialog between two disagreeing spouses the only way to stimulate and achieve growth. *Adam* needed an *ezer knegdo*, a helper to engage in dialogue with him, so that he could realize his true spiritual potential. This would entail risk, but without risk there can be no true progress.

Thus, *Adam* looked for the helper he needed. God showed him all the other animals, but he could not find his mate among them.

We now come to the second problem: what kind of matchmaker fixes up a man with an animal? True, many of us have gone with dates to the zoo, but usually we selected our companions in advance. We didn't go alone to look into the cages and search for that one special mammal who would fulfill our dreams. What was really going on at this stage of man's development? We are not yet ready to answer this question; we must leave it for later in the book. But we will make note of the problem as we write on the board:

26. God brought all the animals before man to see if he could find a mate.

27. Man could not find a partner among the other animals.

Well, we've all had trouble finding our soul mates, but what happened next is something that none of us have experienced. The Torah says that God caused *Adam* to fall into a deep sleep, and He then took flesh from his side and formed from it a second being, a woman. Let us recall that *Adam* was a hermaphrodite, the first human was both male and female. God fissioned this single being into two separate ones, one entirely male and the other entirely female, and they have been trying to get back together ever since.

This is not a joke. From the viewpoint of the Torah, the attraction between a man (now meaning a male) and a woman is the desire to reunite that which was split apart. It is only through a marriage (with dialog) that the unification can occur.

The Torah does not view the separation as having been symmetric. Thus, the being who became the first woman was taken from the side of the original man. The male seems to have gotten the larger share of the body. Following the split, the male is still called *adam* but the woman is called *isha*, "woman," which literally means "from man (male human)." It is *Adam* who gives his mate her personal name, *Chava*, meaning life. It is not within our scope here to discuss these asymmetries, but we do note that man and woman are very different beings. The male is called *Adam*, and somehow he represents all of mankind in a public and open way. The female is called *Chava*, and she embodies all of life, an inner and private quality.

Their natures are different, and their roles are different. This we can understand. Yet there is a final and very noticeable asymmetry that is more puzzling. *Chava* is called the mother of all flesh, but *Adam* is not called the father of all flesh. Weren't they both the original parents of us all? We will discuss this question further in a later chapter, but for now we merely take note of it as we write:

28. God fissioned the original man into separate male and female beings.

29. The female became the mother of all flesh. There is no similar statement concerning the man.

What of the separation itself? In the first creation story we are told that God created man in a manner entirely consistent with the laws of nature, but here we have an event that is quite different. The fissioning of a single person into two violates all known laws of physics and biology. How can we understand this? To say that it was a miracle is insufficient. What do we mean by a miracle?

The Rabbis of the Talmud tell us that there are two classes of miracle. One class is those miracles that occur through the normal laws of nature. We sometimes call them coincidences. An example reportedly took place during the Six Day War.[11] An Arab artillery shell landed in a hospital ward in Jerusalem, but it did not explode. Now, certainly there have been many instances of dud artillery shells that, due to various natural causes, fail to explode. To the occupants of this ward, however, this was a miracle!

Let us notice that the interpretation of this event as miraculous is not forced upon us by the nature of the circumstances. A person could choose to call it an accident, an amazing coincidence. It is only because, in the Torah view of the world, there are no accidents that we identify this as a miracle.

On the other hand, the story of the fissioning of the first person into a man and a woman is an example of the second class of miracle, one occurring outside of the laws of nature.

---

[11]I recall an acquaintance telling me of this incident. I have not been able to verify it. In any case, the reader certainly knows of similar occurrences.

If another person were present observing this event, they would certainly call it miraculous, since the laws of nature do not act in this manner.

Does the existence of miracles violating the laws of nature contradict the scientific viewpoint? Most scientists regard scientific law as immutable and inviolable, and they would therefore say that such miracles are impossible. But a bit of further thought shows that this is not at all a logical position. The nature of science is to infer laws that seem to apply under repeatable and testable circumstance. Science tells us how the world operates when the laws of nature are in effect, it says nothing about what would or could happen if these laws were suspended. The second class of miracle is by definition an event that suspends the laws of nature. Since it suspends them, it does not contradict their applicability at other times. Such a miracle occurs *outside* of the laws of nature; it is therefore outside the domain of science. Since its occurrence means that the laws of nature have been suspended, it can not be contradictory to their existence at other times.

Why are there laws of nature in the first place? After all, the only true mover and shaker is God. To answer, let us remember that the universe is God's hiding place. He must remain hidden so that we can have free will. How does God hide?

To answer this, let us consider why an apple falls from a tree. The apple falls from a tree because God makes it do so. This is the ultimate truth, and there is really little else to say. But, if we perceived God's action so directly, He could not remain hidden. So, He plays a little trick on us. He makes an apple fall down today, and tomorrow, and every single time we drop one. He never makes one fall up. We, as clever observers, get the point. "Aha," we say, "apples always fall down!" We ascribe the falling of apples to a "law of nature."

It is both reasonable and important for us to do this, for we must understand the usual behavior of the world so that we can get on with our lives. We learn to place our apples in baskets rather than attempting to suspend them in mid air. The Torah even exhorts us to observe the regularities of nature and live according to them: we are forbidden from relying on miracles (which violate the laws of nature).

This regularity in God's actions gives us an important choice. If we wish, we can remove God from the picture altogether. We can ascribe the ultimate reality to the laws of nature and even say that there is no God. It is up to us. Therefore, the atheistic scientist can say that physics is the ultimate reality. In his view, violation of the laws of nature is impossible. But the Torah tells us something different. It tells us that, just as God usually chooses to make apples fall down, He could occasionally make one fall up if He wished. Scientists would be astounded. They might even admit that such an event would be miraculous. We, however, see that such an event outside the laws of nature is no more or less miraculous than the regular operation of the laws of physics. They are both due to God's will. It is just that God often acts in a predictable fashion so we can infer "laws of nature" and become atheists if we wish.

To make the above ideas clearer, we will consider an analogy: the dress rehearsal of a stage play. During a dress rehearsal, the actors are in full costume and the stage set is configured just as it will be during the actual performance. Dress rehearsals are almost never interrupted by directors. The actors create a compelling reality upon the stage and, as we watch, we are drawn into their world. It becomes, for us, real and permanent. The director is forgotten.

The curtain opens on act one. It is a dramatic moment. A mother is facing a dreadful choice. She is a Jew trapped with her two small children in Europe during the early days of World War II. She has a chance either to escape on a small boat that will attempt to run the blockade to Israel or to remain in hiding. For half the first act, she puzzles over the decision. We feel very deeply the agony of her dilemma. Finally, she makes her choice. She will leave her hiding place and try to get to the boat. Just as she is about to leave, a parade of soldiers blocks her way. She is forced to abandon her escape and returns to her hiding place. We are deeply depressed. The next day, however, she learns that when the boat tried to leave port it was fired upon, sunk, and all aboard were killed. She screams with sorrow, yet she also exclaims with gratitude: "It was a miracle that we missed the boat. God has saved us!" The curtain falls, and we grip our seats as we await the next act.

After the intermission, the curtain rises. Again, we empathize with the mother's predicament as the play goes on. Soldiers enter the house. She takes the children and hides with them in a closet. All of a sudden, we hear the director speak up. We had forgotten about him completely! He interrupts the rehearsal and tells the stagehands to change the location of the closet. It will make things much more dramatic if the closet is on the left side of the room, near the door where the soldiers are, rather than on the right. The actors wait patiently while the closet is moved. Finally, the stagehands leave, and the play resumes. Very quickly we are drawn back into the world of the play as it reaches its dramatic conclusion. Emotionally drained, we applaud the actors and the author.

Let us consider what we have been observing. In watching the play, we have been dealing with two levels of reality. In truth, this is nothing more than the dress rehearsal of a stage play. But the play is done with such skill that it creates its own world, a second reality that we are drawn into intellectually and emotionally. This little world has laws of its own, and when the mother "miraculously" missed the boat, these laws were not violated in any way. Of course, we know that the author constructed this sequence of events to heighten the dramatic impact; we know it was not an accident. But the mother, within the world of the play, knows nothing of the author or of the higher reality, the theater within which we sit. To her a fantastic coincidence has occurred. It is a miracle within the laws of nature.

In the second act the closet moved from one side of the room to the other. Within the reality of the play, such a thing is impossible. Closets are parts of houses; they don't move. But the closet was not moved within the reality of the play. The play stopped. Its reality was suspended, and in the higher reality of the theater it was stagehands who moved the closet. If a character in the play were to see it happen, he would pronounce it a violation of the laws of nature. Yet, to us, we see that while the illusory world of the play may forbid it, it is entirely consistent with the higher reality of the theater. This is how a nature-violating miracle works. It does violate the illusory laws of our reality, but God is not bound by these laws and can suspend them when and how he chooses.

Incidentally, not only does this analogy help us understand the nature of miracles, but it also sheds light on the classical philosophical problem of free will and God's knowledge of the future. If we indeed have free choice, how is it possible for God to know what we will do?

Consider the play. In the first act, the mother faces an agonizing choice. The play is so well written and acted that we sit on the edge of our seats wondering what she will do. Suppose the play affects us so deeply that we choose to see it again. We are still captivated by the first act. We still feel the tension of her predicament, the agony of her choice. But, just a minute! Don't we know how it will all come out? Haven't we seen the play before? Yes, but if the author is talented, we are so drawn into the play's reality that we forget we are in a theater. Even though in the higher reality of the theater we know what will happen, this does not in any way take away the reality of the mother's free will within the world of the play.

In the reality of our lives, we have absolute free will. But God is not limited to our world. He is akin to the director of the play. He knows what will happen, but this in no way contradicts the reality of our choice within this universe, within the play.

We have come a long way. The blackboard is filled. It is shown in Figure 9-1. We now have a sketch of what the Torah tells us about creation. It is time to consider what the scientists are saying.

NATURAL MAN	CONFRONTED MAN
1. Moment of creation –something from nothing	22. Mist waters earth
2. Continuously renewed creation	23. Man from dust of earth
3. Chaotic nothingness	24. Vegetation
4. Creation through 10 utterances	25. Animals from dust of earth
5. Light	26. Man looks for mate
6. Some light hidden away	27. Man cannot find partner
7. Six twenty-four hour periods	28. Hermaphrodite split apart
8. Concealment of single force	29. Female is mother of all flesh
9. Fluid/vapor gathered together	
10. Force toward one place	
11. First fluid bodies	
12. First solid matter	
13. Genetic material	
14. The heavenly bodies	
15. The sun and moon	
16. Animate life: swarming creatures, birds and fish	
17. Animal soul	
18. Land animals	
19. *Homo sapiens*	
20. Long life span	
21. First human soul	

Figure 9–1: The blackboard

10

Mr. Hubble's Constant
The Expansion of the Universe

When we observed Dr. Selma Wainwright conducting her experiments,[1] we saw a true experimental scientist at work. Not only did she carefully observe the world to formulate her hypotheses, but she actually manipulated and interacted with it by conducting experiments to test her suppositions. She tried out her ideas to see if they were accurate. Medical researchers are not the only scientists who use the scientific method in this classical manner. Physicists, chemists, and biologists are able to formulate hypotheses and conduct experiments to test them. But there are other scientific disciplines in which direct experimentation is not possible. Astronomers can only observe the heavens; they cannot manipulate stars or galaxies to see what happens to them. They cannot try out their ideas to see if they are correct. The same is true for cosmologists, those physicists who study the struc-

---

[1]Chapter 3.

ture of the immense universe as a whole; for cosmogonists, those who theorize concerning the universe's origin; and for paleobiologists, who study the origins and evolution of life on earth. The universe is too big and the origins of life are too far in the distant past for them to poke, tweak, and manipulate to see what happens.

Researchers dealing with the beginnings of life or the origin of the universe are engaged in what we might call extrapolative science.[2] That is, they are observing the present state of the world and attempting to develop theories that can be projected backward in time to explain events of the distant past. The limitations of this endeavor are obvious. Since they can only observe, their theories are necessarily less verifiable than those of experimental physicists or biologists. Their conclusions are less precise and more subject to error.

But this does not mean that cosmogony or paleobiology are foolish pursuits. The researchers are merely doing the best they can. As we pointed out in Chapter 3, although they cannot conduct truly repeatable experiments, they can make predictions and perform observations to verify their hypotheses. For example, an evolutionary biologist might theorize that life evolved slowly from more primitive forms. Based upon this hypothesis, he would predict that as he dug deeper and deeper into the earth and uncovered older and older layers of rock, he would find fossils of animals that become increasingly primitive. He could then dig, and see what he found. This digging into the earth would be the closest he could come to actual experimentation.

---

[2]M. M. Schneerson, "A Letter on Science and Judaism," in *Challenge*, ed. Aryeh Carmell and Cyril Domb (New York: Feldheim, 1976), pp. 142–149.

As long as we understand the inherent limitations in extrapolative science, we can study and appreciate it for what it is: the closest we can come to a truly scientific study of the past. It is not perfect, but it is the best any of us can do to develop an understanding of how what is came to be.

When we discuss the scientific picture of the creation, we will be discussing the results of extrapolative science. Since it will be important to keep the above limitations in mind, we will begin with an actual case history of the development of an extrapolative scientific theory. We will pick one of extreme interest to us. This theory will turn out to be of major importance in determining just how the world came to be.

The theory we will consider arises from the consideration of a simple question: how big is the universe? This is an easy question to ask, but not so easy to answer. How do astronomers ever know how far away anything is? It turns out that there is no single answer. The distances to different things are measured in different ways. Let us start with the way in which astronomers determine the distance from the earth to the sun.

We have all had the annoying experience of sitting in a movie theater[3] and finding a rather tall person in the seat directly ahead of us. We end up constantly moving our heads from side to side to get a good view of the movie. Why don't we just keep our heads fixed? Because when we move to the left, the fellow's head blocks the right side of the screen, and when we move to the right, it is just the other way around. Moving our head is a bit of a pain, but it works. What, however, if there were a basketball player sitting not in the seat directly in front of us but several rows ahead? Just moving

---

[3]Some readers may wish to substitute *"bais medresh"* for "movie theater."

our heads from side to side would no longer be enough. We would have to get up and move several seats over to get the same effect. The further the tall person is from us, the further we would have to move.

All of this is an example of what astronomers call parallax. Parallax is the apparent shift of an object against a background that occurs when we move from side to side. In the movie theater, it is parallax that causes the head in front of us to appear to move to the left of the screen when we move right and vice versa. The closer something is to us, the greater is the effect of parallax. That is why we can simply move our heads a little to see around the fellow in the seat ahead, but we must make a greater effort to see around someone a greater distance away.

Now, suppose we observe the sun against the background of the stars. This is a bit tricky to do, since we cannot see the stars in the daytime, but it can be done during a solar eclipse. Suppose that while observing the sun we move from side to side. The position of the sun would appear to shift relative to the background of stars; if we move to the left the sun appears to shift to the right and so on, just as in the movie theater. Of course, the sun is pretty far away, so we would have to move quite a bit to make it noticeable. A simple way to do this is to merely stay put and allow the earth to carry us from one side to the other as it rotates on its axis. This moves our position by about eight thousand miles, the diameter of the earth. Let's recall that the closer an object is, the greater will be the parallax, the apparent shift. By knowing how far we have moved from side to side, a distance we call the baseline, and measuring the amount of the parallax, we can tell how far away the sun is.

Actually, the measurement of the distance to the sun is a little more complicated than this. It was originally done indirectly using the parallax of a small asteroid called Eros[4] and a whole bunch of equations relating Eros' distance to that of the sun, but the concept we have presented is basically correct. The answer comes out to be about ninety-three million miles (150 million kilometers). This distance is so important in astronomy that it is given a name: the astronomical unit.

Well, we now know how far the sun is. The next step is to determine the distance to some nearby stars. Can we do this by measuring their parallax? The stars are much farther away than the sun. Therefore, their apparent shifts are much smaller. In the movie theater, it was not sufficient to merely move our heads to see around the person several rows in front; we had to actually get up and move to another seat. With the stars, the eight-thousand-mile baseline that we get by riding on the earth for twelve hours is not enough; we must find a way of moving a greater distance. The solution is to ride on the earth not for twelve hours, but for six months. This is half the time it takes the earth to go completely around the sun, so we end up on the opposite side of our solar orbit. Our baseline is then twice the distance from the earth to the sun, two astronomical units, or about 186 million miles. Now, when we observe the stars, we see that many of them do appear to shift to the left or the right every six months. We can measure their parallaxes and determine how far away they are.

The closest star to the earth turns out to be Proxima Centauri, which is about 25,000,000,000,000 miles away. This is a very big number. Since we will be talking often about very large or very small numbers, we must get used to writ-

---

[4]O. Struve, *Elementary Astronomy* (New York: Oxford University Press, 1959), p. 62.

ing them the way scientists do, in terms of powers of ten. By a power of ten, we mean a number such as 10^N which means 1 followed by N zeros. That is, we write 15 as 1.5×10^1 (that is, $1.5 \times 10 = 15$), 3,000 as 3.0×10^3 (3×1000) and 25,000,000,000,000 as 2.5×10^{13} ($2.5 \times 10,000,000,000,000$). When it comes to very small numbers, we write them similarly using negative powers of ten, where 10^{-N} means 1 divided by 10^N. Thus, 0.015 is written as 1.5×10^{-2} ($1.5 \times .01 = 1.5 \times 1/100$), 0.0000003 as 3.0×10^{-7} ($3.0 \times 0.0000001 = 3.0 \times 1/10,000,000$) and so on.

Let us recall that light moves at approximately 186,283 miles per second (1.86283×10^5). In a year, there are approximately 31,557,600 seconds (3.15576×10^7). Therefore, in one year a beam of light would travel $186,283 \times 31,557,600 = 5,878,644,401,000$ miles (approximately 5.8786×10^{12} miles). This distance is called a *light year*. It is a measure of distance, not time.

Let's get back to Proxima Centauri. We saw that it is 2.5×10^{13} miles away. From the numbers above, we can now calculate that it is 4.3 light years from the earth. The astute reader will have noticed something a little unsettling. To determine the distance to this star, astronomers not only had to measure its parallax, but they also had to know in advance the size of the baseline, twice distance from the earth to the sun. This distance was itself determined through observation of the sun's parallax. Suppose there were some error in the determination of the distance from the earth to the sun. Then this would also cause an error in the determination of the distance to Proxima Centauri and all other stars. This is typical of what happens in extrapolative science. The conclusions become increasingly indirect, increasingly dependent upon other results, which may themselves be subject to errors. Again, this is not bad science; it is just the best that can be done.

We saw that Proxima Centauri was quite far away, yet it is the closest star to the solar system. Proxima Centauri and the sun are only two stars among the billions that make up our galaxy, the huge aggregation of stars we call the Milky Way. Our galaxy is, in turn, only one of billions of galaxies in the universe. Within our galaxy there are many stars that are so far away that their parallaxes are too small for us to measure without a larger baseline. One way to get one is to wait patiently while the entire solar system moves through space. In a period of ten years, the solar system moves about forty astronomical units through the galaxy. This gives astronomers the much longer baseline they need to work with. But there is a problem. Over a ten-year period, the stars also move a bit on their own. This creates a very difficult geometry problem, but astronomers are paid to deal with tough problems. Over the years, they have found ways of solving this problem,[5] and the distances of stars in the far reaches of the galaxy have been determined.

What about stars that are still further away, such as those in other galaxies? Is there any way we can determine their distance? The answer is yes, but to understand how we must first take a little detour and discuss the fascinating subject of variable stars.

Unlike our sun, whose brightness is fairly constant from day to day, there are many stars that pulsate, becoming alternately brighter or dimmer every few days. It would be pretty annoying if the sun did this, but the existence of variable stars gives astronomers plenty of work to do. They have studied them exhaustively and found that there are several distinct types. One particularly interesting class is called Cepheid vari-

---

[5]Actually, through statistical techniques. See Struve, section 24.7.

ables. They can be easily recognized by the patterns of their pulsations.

In 1912, an astronomer named H. Leavitt became interested in the Cepheid variables. In particular, she wondered how their pulsations were related to their average brightness. Now, when we speak about how bright a star is, we have to be careful. Consider a lit candle. If we are standing close to it, it can seem to be quite bright. But from a few hundred feet away, it would seem pretty dim. If it were a mile away we would not be able to see it at all. The intrinsic brightness of the candle doesn't change, but the apparent brightness does. To determine how bright something really is, we must measure both its apparent brightness and its distance and use the appropriate formula to make the correction. An astronomer named Harlow Shapely did this for the Cepheid variable stars in our galaxy. He measured their apparent brightnesses and determined their distances by the parallax method discussed above.

Together, Leavitt and Shapely discovered something quite important about Cepheid variables.[6] They found that the brighter they were, the more slowly they pulsated. Not only that, but there was a definite predictable relationship between their brightness and the period of their pulsations. Thus, if they found a Cepheid variable and didn't know how far away it was, they could still determine its intrinsic brightness just by observing the period of its pulsation.

This discovery gave astronomers the yardstick they needed to measure the distances to other galaxies. It works as follows. First, they look at the stars in another galaxy until they find a Cepheid variable. Then, they observe it for a while

---

[6]Struve, section 30.8.

to determine its period of pulsation. Once they know this, they can compute its intrinsic brightness, since this is determined by its period. Next comes the big step. Since they now know the intrinsic brightness of the star, and since they can measure its apparent brightness, they can compute how far away it is.

Well, astronomers were now able to measure the distances to some nearby galaxies. But we all know what is coming next! Many (indeed most) galaxies are too far away for us to be able to see individual stars within them. Some are so far away that they look like faint little smudges of light. How can we measure the distance to the furthest galaxies?

To understand how this is done, we must take another detour. Let's think about the sound an airplane makes as it screeches by overhead. When it is coming toward us, the sound is fairly high pitched, but when it is going away, the pitch is lower. Every 10-year-old boy knows this. We only have to listen to the sounds he makes when he plays with his jet fighters: "eeeeeoooooouuuuuu!" Physicists call this the Doppler effect. We will not get into an explanation of why it happens, but the Doppler effect causes the apparent frequency of the sounds emitted by an object to be higher when it is coming toward us and lower when it is moving away.

This doesn't only happen with sound. It happens with any wave phenomenon. Let's recall that light also acts as a wave. It has a frequency of oscillation, and this frequency is higher for blue light and lower for red. Since light is a wave, the light waves emitted by a moving object are subject to the Doppler effect. This means that if the object is moving toward us, the apparent frequency of the light will be increased and the color of the light will be slightly bluer than if the object were stationary. Conversely, if the object is moving away, the color of its light will be shifted slightly toward the red.

In daily life we do not notice this because the effect is small. An object would have to move tens of thousands of miles per second before we would be able to perceive the Doppler shift in its color with our naked eyes. But sensitive instruments have verified the reality of this effect, and it is used quite effectively by police radars to determine how fast our car is going.

Back to the distant galaxies. Armed with the Cepheid variable method of measuring distances, astronomers had a field day measuring the distances to a whole bunch of nearby galaxies. One of the astronomers involved in the research was a man named Edwin Hubble. He began to notice something very interesting.

He noticed that the light from the nearby galaxies tended to be a little redder than that from our galaxy. What was the cause of this shift toward the red? There were several explanations that he considered. One, for example, was the possibility that there was something in space, such as huge clouds of gas, that absorbed blue light. If this were true, a galaxy that was viewed through one of these clouds would have its blue light partially absorbed and thus would look reddish. This would also explain why some galaxies looked redder than others. If a cloud of gas happened to be in the way, it would look redder; otherwise it wouldn't.

If the red shift were caused by clouds of gas, galaxies that appeared close together in the sky should appear equally red, since their light would tend to pass through the same gas clouds when reaching the earth. When Hubble searched the sky, he found that this was not so. Often, a galaxy with a pronounced red shift would be right next to one with a much smaller red shift. Hubble abandoned this hypothesis and looked to see if he could find any other patterns in the reddening. This led him to his greatest discovery.

Hubble looked at the distances to the various galaxies as determined by observations of Cepheid variables. He noticed that the further away a galaxy is, the greater is its red shift. The relationship turns out to be quite simple and predictable. If one galaxy is twice as far as another, it tends to have a red shift that is twice as large. This gave him an idea. Suppose the red shift is caused by the Doppler effect. Then, galaxies that are moving away from the earth would have their light shifted towards the red by an amount dependent upon their velocity. What about two galaxies that appear to be close together in the sky? They could be moving away from the earth at very different velocities, and they would then have quite different red shifts.

To summarize Hubble's discovery: the galaxies of the universe are receding from ours at various velocities. The further away they are, the more rapidly they are moving away from us. Because of this motion, the Doppler effect causes the light from a galaxy to be shifted towards the red. Therefore, the further away a galaxy is, the greater its red shift.

The pattern of galaxies' motion is so regular that Hubble was able to predict how fast a galaxy would be receding from us if he could measure its distance. He found that, on the average, galaxies a million light years distant are moving away from us at about twenty-six kilometers per second, those two million light years away are receding at fifty-two kilometers per second and so on. The number that summarizes this pattern, twenty-six kilometers per second for each million light years, is called Hubble's constant.

Hubble's research dealt only with the closest galaxies, those whose distance could be measured using Cepheid variables. But Hubble and other astronomers couldn't resist the temptation to generalize this discovery to all galaxies. They formulated Hubble's law, which states that *all* galaxies are receding from ours at a rate of twenty-six kilometers per second for each million light years of distance. How could astronomer's take this giant leap in generalizing Hubble's research? Well, scientists do things like this all the time. It's not bad science, as long as they carefully state their hypotheses and test them whenever they are able. To date, Hubble's law has held up pretty well.

Hubble's law gave astronomers the means they sought to determine the distances to all galaxies in the universe. They just had to reverse the reasoning of Hubble. To see how this works, let's consider a very faint and apparently distant galaxy. Astronomers observe its light and find that it is shifted strongly toward the red. From the magnitude of its red shift, they can use the Doppler effect formula to determine how rapidly it is moving away from us. Let us suppose they find it to be receding at twenty-six hundred kilometers per second. Since Hubble's constant implies that galaxies recede at twenty-six kilometers per second for every million light years of distance, this particular galaxy must be one hundred million light years away.

That's pretty far, but astronomers have detected galaxies billions of light years away (and remember, one light year is about 5.9 trillion miles). The universe is a pretty large place, and all those galaxies are moving away from little us! Does this make sense? Aren't there any coming the other way? Does this mean that there is something special about our place in the universe?

As we might expect, scientists do not wish to ascribe any special significance to our location. They assume that every place in the universe is pretty much the same as any other. Of course, they give an impressive-sounding name to this idea, they call it the *homogeneity of space*. So how can all the galaxies be receding from a single point if that point is not special in any way?

To answer this, let us consider the surface of a small balloon. Before blowing up the balloon, we paint small dots on it. On the average, the dots are about a quarter of an inch apart. Now we blow up the balloon. Suppose it increases in size by a factor of four. The dots will now be about an inch apart on the average. If we were sitting on one of the dots, we would see that all of the other dots got further away. Yet, there is nothing special about any one dot. All the dots are getting further away from all the other dots.

Astronomers now think that the galaxies in the universe are similar to the dots on the balloon. They are all getting further away from one another. The reason that this is happening is because *the entire universe is expanding*! There is a somewhat amusing story connected with this.

Hubble made his discovery in 1929. A dozen years earlier, there was another young physicist named Albert Einstein who was working out a theory that came to be called *general relativity*. (We will talk a bit more about Einstein in a later chapter.) Einstein's theory explained how gravity worked, and since gravity is the most important force in determining the behavior of large objects, physicists were able to predict the behavior of the universe as a whole. This was well and good, but there was one problem. The darn thing just wouldn't hold still! No matter what assumptions they made about the size and total mass in the universe, it turned out that it would always end up either expanding or contracting. But in 1917, everyone "knew" that the universe didn't do this. All physicists believed that the universe was fixed in its size. Einstein was also bothered by this conclusion, so he tried to fix his theory by adjusting his equations. He added a number he called the cosmic constant. By choosing an appropriate value for this constant, he could get the universe to stand still. Then came 1929, and Hubble's discovery that the universe was expanding, just as Einstein's original theory had predicted. Einstein later said that the adjustment of his original theory was "the biggest blunder of my career."

Let's summarize what we have learned. Astronomers now believe that the universe is expanding. This knowledge is based upon Hubble's law, which is a generalization of Hubble's constant, which is determined through observation of Cepheid variable stars, which are studied through parallax measurements, which are possible because we know the distance from the earth to the sun, which we know through other parallax measurements. . . . It's all beginning to sound like one of those songs we sing at the end of the Passover *seder*. There is a long chain of indirect reasoning. It is a little like a house of cards: take one away and they all fall.

Again, *this does not make it bad science*! It is the best we can do. In fact, it is actually a bit better than a house of cards because there are some other ways of independently verifying some of the conclusions. For example, galactic distances can also be measured by using the brightness of certain types of galaxies, just as astronomers use the brightness of Cepheid variable stars to determine their distances. But remember that we are dealing with extrapolative science. Cosmology can never be as precise or testable as experimental chemistry.

We have taken up the subject matter of this chapter for two reasons. The first was to show just how cosmological science is practiced, to show how tentative and uncertain its conclusions are even when done by the most brilliant and careful scientific minds that have ever lived. We must therefore not be surprised if we read one day that scientists have (again) radically revised their theories of the entire universe. For example, even though we have used the figure of twenty-six kilometers per second per million light years as the value for Hubble's constant, some astronomers[7] now calculate a number as large as thirty-two or as small as sixteen. Even a small change in this number changes the computed size of the universe by hundreds of millions of light years. All science is tentative, and some sciences are much more tentative than others. We must not forget this.

The second reason we have brought all this up is to introduce the expansion of the universe. This turns out to be very important for scientist's current understanding of the origin of the universe, which we will discuss in the next chapter.

---

[7] "Hubble Constant," *McGraw Hill Encyclopedia of Science, 7th Edition*, ed. Sybil Parker et. al. (New York: McGraw Hill, 1992), 8:517–518.

11

Bang
The Origin of the Universe

Up to now we have stressed the concern of Torah with what *ought to be*, and the concern of science with what *is*. But the scientist is not so easily satisfied with what *is now*; he is also driven to ask *what was* and *what will be*. In his quest to understand the physical world, he cannot help but wonder where it all came from and what its ultimate fate will be.

The past or future cannot be observed in the laboratory. The scientist must, therefore, observe the present world and attempt to extrapolate what he sees backward into the past or project it forward into the future. This is not the way he would like it, but it is the best he can do. To guide him in this pursuit, he makes use of an important principle called Occam's razor. This rule was first formulated by William of Occam (c. 1285–c. 1349), who put it this way: "What can be done with fewer [assumptions] is done in vain with more."[1]

---

[1] "William of Occam," in *The New Columbia Encyclopedia,* ed. William H. Harris and Judith Levey (New York: Columbia University Press, 1975) p. 2981. Reprinted with permission of the publisher.

In modern terms, this means: if we have two different theories and they both successfully explain all the facts, we choose the one that is simpler. The principle is called a razor because it is used to cut out the more complicated explanation.

To make this more understandable, let us consider an example. We are watching television, and we see the interior of a small room. Suddenly, we notice a small steel ball moving through the air. It hits the floor, bounces a few times, and then remains in place. What has happened? There are at least two theories that we can use to explain what we have seen. The first is that we are looking at a normal room and gravity caused the ball to fall down and come to rest on the floor. This is, of course, an obvious explanation, but it is not the only one possible. Perhaps the room and camera are actually upside down. Perhaps when the ball moved towards the floor, it was really moving upward. Why would it do this? Because, unseen by us, someone put a giant magnet on the other side of the floor, and it is pulling the steel ball in this direction.

"Oh, come now," a person might be tempted to say, "give me a break! Do you expect me to believe that?" If he feels like saying this, he is unconsciously applying Occam's razor. He is actually saying, "Why go to such convoluted lengths to interpret what we have seen when there is a much simpler explanation?"

Occam's razor is basic to all science. How, for example, does a scientist know that the laws of physics are the same today as they were a billion years ago? After all, the physicist cannot prove that the laws of nature have not changed. He cannot go backward in a time machine to measure the strength of gravity as it was a billion years ago. Maybe gravity was stronger then. In fact, there is a very controversial theory due to P.A.M. Dirac, one of the greatest physicists of the twentieth century, that implies that the law of gravity does change with time. This change would be so slow that it would not be possible for us to observe it directly with current scientific instruments. Thus, such a variation is indeed possible. It's just that most scientists apply Occam's razor and say, "Why go to such great lengths to make a complicated theory when we can explain all present experiments with the simpler assumption that the laws of physics do *not* change!"

From a Torah viewpoint, we would say that Occam's razor leads the physicist to say, "God made a leaf fall yesterday, and He made it fall today. I guess He'll make it fall tomorrow, too." There is no logical reason that this must be so. It is just that there is no reason to assume anything more complicated. In the words of the gambler: "The race is not always won by the swift, nor the contest by the strong, but that's the way to bet!"

Let us now return to the study of the universe's origin. Armed with Occam's razor, cosmologists began to investigate the nature of the cosmos. They observed planets, stars, and galaxies. The simplest thing they could say was: "This is the way the world is now; so must it always have been." This assumption, that the universe has always been pretty much as it is now, is called the *steady state theory*. It is intuitively satisfying; it is simple; and it explains a lot. But it turns out that there is a very subtle problem with it, and it all has to do with air conditioning.

To understand the relationship between air conditioning and creation, we must first take a little detour. We must introduce a fascinating concept that physicists call *entropy*.

Entropy is a way scientists characterize the chaos or disorder in the physical world. As an example, let's look at that small patch of grass in front of our house. A neatly manicured lawn is orderly, and it has a low entropy. A chaotic, overgrown one is disordered, and it therefore has a much higher entropy.

Physicists can define entropy mathematically so that it is an exactly measurable quantity.

The amount of entropy in the universe is not constant. It increases in time. Thus, the universe is continually becoming less and less orderly. The fact that this happens is given a rather impressive title; it is called *the second law of thermodynamics*. Let's think again about the lawn. Left alone, the neat lawn will not remain well manicured. In a rather short time, it will become an overgrown mess. This is an inescapable consequence of the increase of entropy with time. The only way to prevent the lawn from progressing into chaos is to expend some energy to get it back into shape. Although the second law of thermodynamics does state that the entropy of the universe as a whole must always increase, we can decrease the entropy of a small piece of the universe if we want to, but we must expend energy to do this. Thus, to make our lawn neater, we must run our lawn mower. We do this by using the chemical energy stored in gasoline. When we run out of gas, we have to buy some more. But, sooner or later, the universe will run out of gas, and the lawn will win.

The fact that we cannot find a limitless source of energy is a consequence of another physical law, *the first law of thermodynamics*. This states that the total amount of energy in the universe is fixed. We can transform one type of energy into another, but we cannot make any more than there is already. This is also called the conservation of energy.

Now that we have discussed the first two laws of thermodynamics, we might wonder what this word, *thermodynamics*, means. Since it begins with the prefix *thermo* (as in thermometer and thermos bottle), it is apparent that it has something to do with heat. Thermodynamics is the science of heat. It describes the nature of heat, how it is produced, and how it behaves. It also tells us that heat is a form of energy.

The first law of thermodynamics (conservation of energy) tells us that heat can be converted into other forms of energy and vice versa. When we burn some coal, the chemical energy within it is converted into heat. When we burn this coal in an electrical power station, the resulting heat energy is converted to electrical energy. But we don't get something for nothing, every bit of energy must come from some other form of energy. The total amount of energy is conserved.

The second law of thermodynamics tells us that the disorder of the universe always increases. How does this relate to heat? The atoms and molecules that make up all substances in the world are in constant motion. They are always vibrating or moving around in a chaotic fashion. When the temperature of a substance increases, this motion becomes more disordered, and the entropy of the substance also increases. The second law of thermodynamics is equivalent to the observation that, no matter what we do in this world, we always end up producing a little heat, thereby increasing the disorder in the universe.

If we go out for a morning jog, we come back hot and sweaty because, in the process of converting stored food energy into muscular energy, our bodies necessarily produce some heat. Electrical power stations use large cooling towers to dissipate the waste heat they produce while converting the energy in their fuel to electrical power. Notice that we call this "waste heat" because it represents a loss of energy. Not all of the energy in the coal can be converted into electricity. The second law of thermodynamics requires that some of it must be wasted and converted to heat.

In colloquial terms, the first of thermodynamics laws tells us, "you can't win"; the second says, "you can't even break even."

Suppose we watch a movie in which a bomb explodes. We then watch the same movie shown backwards. We can always tell the difference. Why? When a bomb explodes, the chemical energy stored within it is converted to mechanical energy (the motion of the fragments), light, and, of course, heat. A very ordered state is replaced by a chaotic disordered state. Entropy has increased. What if the movie were reversed? The disordered state of fragments, gasses, and heat is seen to spontaneously coalesce into a very ordered bomb. Now, this does not violate the first law of thermodynamics since the total energy is conserved, but it certainly does violate intuition. Things like that just don't happen in nature. Entropy cannot spontaneously decrease. It would violate the second law of thermodynamics. It is the behavior of entropy that enables us to tell whether the movies are going forward or backward. Physicists say that the second law of thermodynamics provides the "arrow of time."

Now let's get back to our air conditioner. The purpose of the air conditioner is to cool the room. To do this, it must remove heat energy. Energy cannot be created or destroyed, so where does it go? Outside. That's why room air conditioners are always placed in windows. Technically, the air conditioner is a "heat pump." It pumps heat from the house into the surrounding air.

Before we turn it on, the room and outside air are both at the same temperature. With no particular pattern of hot and cold, this is a fairly disordered high entropy state (well—this might not be obvious, but the mathematics shows that it is so). We want to produce a state in which the room has gotten cooler and the heat has been pumped from the room into the surrounding air, making the outside a little bit warmer in the process. This will produce an organized pattern: cool inside, warm outside. Thus, the house and its surroundings become a little bit more ordered and their total entropy decreases (again, mathematics proves that this is true). It takes energy to do this. That's why we must plug the air conditioner into the electrical outlet and pay those large bills to the electric company.

But what of the second law of thermodynamics? It states that entropy must increase. How could we make it decrease? The answer is that the second law actually applies to the universe as a whole. Entropy can decrease in one place as long as it increases somewhere else in such a way that the net effect is an increase in the total. Our air conditioner decreased the entropy in our neighborhood but the electricity we used came from the power station, and in generating this power the station produced a whole lot of waste heat. This waste heat represented a very large increase in entropy that more than offset the puny decrease in entropy at our house.

So how does all of this create a problem for the steady state theory of the universe? (Remember that?) It has to do with what happens when we turn our air conditioner off. In a fairly short time, the pattern we have created, cool inside/warm outside, disappears. The room warms up to the outside temperature, and the outside gets very slightly cooler. The ordered state disappears, entropy increases, and everything approaches the same temperature. Without an input of energy, entropy increases and everything becomes disordered.

The universe as a whole does not have any input of energy. None can be created; this is the first law of thermodynamics. Therefore, its entropy inexorably increases. Oh, we can occasionally create little pockets of decreased entropy by connecting our air conditioners to power stations and moving energy around, but we can never prevent the overall march toward chaos and disorder. Just as our local order disappears without an energy source and everything approaches the same temperature, eventually the universe as a whole must proceed to a state of maximal disorder. The entire universe must ultimately end up in a chaotic state in which everything has the same temperature. Everything will decay and fall apart. There will be no stars, no galaxies, no living organisms, no hamburgers with fries. There will just be a uniform and very boring gaseous soup filling all space. Nothing can prevent this from happening. Physicists call it the "heat death" of the universe.

Now the problem with the steady state theory of the universe becomes clear. If the universe has always existed, that is, if it has been around for an infinite length of time, then it would already have reached the state of heat death. But we are all here to read these words. We have hands, eyes, and brains. We are ordered beings in a partially ordered universe. Temperatures vary. The universe is young enough for all this to be true. If the universe had been around for ever, the existence of this order would violate the second law of thermodynamics.

Many physicists were deeply bothered by this. They tried many ways to get around the problem, but clever as they were, they soon faced an even bigger objection to the steady state theory.

According to this theory, the universe must always have been just as it is now. But Edwin Hubble's observations showed that the universe is expanding. All galaxies are flying apart at thousands to hundreds of thousands of miles per second. It is almost as if we are looking into the sky just after the burst of a huge Fourth of July firework. We see colored sparks flying in all directions, yet all of the debris originates from a single explosion.

If we look at the positions and velocities of all of Mr. Hubble's galaxies and extrapolate backward in time, we find that some ten to twenty billion years ago they would all have originated from the same point. The universe looks as if a major explosion took place at that time and that all the matter in the universe is no more than debris from this explosion. All the planets, stars, galaxies, and clusters of galaxies originated from this burst, and they have been flying apart ever since. This explosion is called the "big bang."

The modern theory of the big bang originated in 1946 in a paper published by the physicist George Gamow.[2] Gamow extrapolated the current state of the universe backward in time to the first few fractions of a second following the big bang. With so much matter in one place, the conditions were quite a bit different than they are today.

Let's imagine that the entire earth, which has a mass of about 1.3×10^{26} (130,000,000,000,000,000,000,000,000) pounds, was squeezed into a soda can. This would be a pretty hard thing to do, and we would end up with a pretty heavy can. The material in the can would be about 1.8×10^{25} times denser than water. Yet this is nothing compared to the density that existed when all the matter in the universe was compressed into a volume smaller than a poppy seed. It is almost impossible for us to imagine what the conditions of the universe were in the first seconds of its existence, but clearly things were much different than they are now. If the big bang theory is correct, it directly contradicts the steady state picture of the universe.

The steady state theorists were not so willing to let their viewpoint go. Some objected, on philosophical grounds, to the idea that there was a moment of creation. How, then, could they account for the expansion of the universe? They

---

[2] G. Gamow, "Expanding Universe and the Origin of the Elements," *The Physical Review* 70 (1946): 572–573.

came up with a very radical and clever idea.[3] Suppose the first law of thermodynamics, the conservation of energy, was not really correct. Suppose that energy could be created spontaneously out of nothing. Since Einstein showed that matter and energy are interchangeable (remember his famous formula $E = mc^2$), this would mean that matter could suddenly appear at random anywhere in the universe. Of course, no physicist has ever observed this (that is why the conservation of energy was formulated as a physical law in the first place), but they theorized that the creation of matter was so infrequent that it would be highly unlikely that even one electron would spontaneously appear on the entire surface of the earth in the history of civilization. Therefore, the conservation of energy would still be approximately true, and the actual violation of it would be very hard to observe.

How could this daring hypothesis save the steady state theory? Let's recall the analogy, in the last chapter, of the dotted balloon. Originally the balloon had dots that were a quarter-inch apart. As we blew it up, the spacing of the dots increased. Suppose that whenever two dots got to be a half-inch apart, we painted another dot half way between them. As we blew up the balloon, we continued to paint dots. Even though the original dots were getting further and further apart, the new dots filled in the spaces.

This is what the steady state theorists proposed for the universe as a whole. As the galaxies move further apart, matter would be spontaneously created to fill in the intervening space. This matter would accumulate into huge clouds of gas that, after a few billion years, would condense into new galaxies.

---

[3]D. A. Evans, *Mathematical Cosmology* (Oxford: Clarendon Press, 1977) pp. 123–129.

Now, if the concept of a moment of creation was hard for secular physicists to take, we can imagine the resistance to this idea of constant creation of something from nothing. Yet good physicists shouldn't let their emotions get in the way of their science. As unconventional as this idea was, it was worthy of serious study, and it raised an important question. Was the steady state theory correct, and was the universe fifteen billion years ago pretty much the same as it is now; or was the big bang theory correct, and was the universe very different then? How could this question be answered? If only there were some way that we could look backward in time to see what was really going on.

Well, there is. We don't know how to build time machines, but we do have something almost as good—ordinary telescopes. Because the speed of light is finite, it takes time for the light to travel from a star toward the earth. The nearest star is the sun, and it takes about eight minutes for light to traverse the ninety-three million miles from it to the earth. Therefore, when we view the sun, we are really looking at things that happened eight minutes ago. The brightest star in the sky is Sirius, which is 8.8 light years away. Sirius is actually a binary or double star. What appears to the naked eye to be a single star is actually two stars very close together. They orbit around one another making a complete circuit every fifty years. But when an astronomer observes their relative position, he is not seeing them where they are now, but rather where they were 8.8 years ago. He is actually looking back into time. Similarly, as we look at objects further and further away, we are looking further and further into the distant past.

In the year 1054, Chinese and Japanese astronomers reported the sudden appearance of a new star that was so bright that it could be seen even during the daytime. Over a two-year period, it gradually faded from view. We now know that what they were observing was a supernova, the very rare and extremely violent explosion of a star. The remnants of this explosion are still visible, a rapidly expanding cloud of gas called the Crab Nebula. This nebula is about forty-three hundred light years away, so the explosion observed by the Japanese and Chinese in 1054 CE actually took place much earlier, roughly in the year 3250 BCE.

The furthest object visible with the naked eye is called Messier 31. It is a large spiral galaxy that appears as a hazy smudge in the Andromeda constellation. When we step out of our houses to view the Andromeda Galaxy, we are looking about 1.4 million years back into the past. But with telescopes, we can see much further back in time. We can observe galaxies that are billions of light years away, and can consequently observe the universe as it was when it was much younger than today.

In the 1960s, astronomers observed a new type of object in the sky. At first, these objects seemed to be merely unusual stars within our galaxy. But when astronomers looked at their colors, they found them to be shifted towards the red far more than any other objects they had ever observed. These red shifts indicated that these objects were moving away from the earth at velocities up to eighty percent of the speed of light. Because the first ones discovered were powerful sources of radio waves, scientists called them quasi-stellar radio sources, or quasars. All quasars are moving away from us at immensely large speeds; none are coming in the other direction. Astronomers now believe that their recessional velocities are due to Hubble's law. They are so far away from us, further than the furthest visible galaxies, that the overall expansion of the universe is causing them to recede at speeds greater than anything we have ever seen before.

Quasars are very mysterious. They are so far away that, to be visible from earth, they must be extremely bright. Most quasars are hundreds of times brighter than entire galaxies. Yet they are much smaller. What are they? We do not know. We know that something very violent goes on inside them, but what it is we cannot say.

What we do know is that there are no quasars around our part of the universe. They are all very distant. Actually, they are the most distant objects we can see. But let's remember that when we look at the distant reaches of the universe, we are actually looking backward in time. Therefore, it is more accurate to say that there are no quasars around *today*, and that there haven't been any for the last 8 billion years.

The discovery of quasars dealt a death blow to the steady state theory. If the universe was always as it is today, why would we see it filled with mysterious quasars a mere 8 billion years ago when there are none to be found at our time? Clearly, the universe has changed drastically since then. It changes; it evolves. The existence of quasars (and other evidence that we will discuss in Chapter 17) has overthrown the steady state theory. The big bang is now almost universally accepted as the origin of the universe. In the words of David Schramm, "We are no longer discussing what the basic cosmological model should be—Big Bang or Steady State. We are concerned now with working out the details of our Big Bang model."[4]

Physicists now speak of a standard model of the universe's origin. Most believe that they can extrapolate backward to understand what happened a small fraction of a second after

---

[4]D. Schramm, "The Early Universe and High-Energy Physics," *Physics Today* 36, no. 4 (April 1983): 27–33.

the big bang. The standard model extends knowledge of the universe back to 10^{-35} seconds after the creation, when all the matter of the visible universe was still scrunched together at incredibly high density. With all of that energy concentrated together, the universe was very hot. The temperature, about 10^{28} degrees, was so high that matter as we know it would be torn apart. Physicists cannot reproduce these conditions of density and temperature in the laboratory. Therefore, for the most part, they must extend current theories to attempt to understand what was going on. But there are a few experiments that can be done that approximate, at least a little bit, what was going on in the early history of the universe.

Physicists can accelerate tiny subatomic particles until they have very large energies, energies almost as large as those present in the first fraction of a second after creation. They conduct experiments in which they allow these particles to collide with each other, and they observe the debris produced by these collisions. A lot can be learned about the basic laws of the universe in this manner. This is the reason that many scientists wanted the United States Government to spend billions of dollars to build, in Texas, a machine 50 miles in diameter that would have been called the superconducting supercollider. If this machine had been built, it would have allowed physicists to duplicate conditions that existed even earlier than 10^{-35} seconds after creation.

Ultimately, how far back in time will scientists be able to go in their reconstruction of the universe's origin? Some cosmologists are talking about a new theory called *cosmic inflation* that could explain what happened when the universe was only 10^{-45} seconds old and the temperature was 10^{32} degrees. But as they go back further and further toward the beginning, the densities and energies become so high that it will never be possible to even approximate them on earth. The origins of the universe will therefore be forever masked in uncertainty as the validity of any physical law we can conceive of becomes more and more speculative.

The standard model can take us back to an epoch in which the laws of physics as we know them are no longer valid, but this doesn't mean that some other laws couldn't be found to explain what was happening. Just because we don't *know* what laws were in effect doesn't mean that physics was not in some way operative at this time.

But the standard model is only a theory of what happened in the short time following the big bang. What about the moment of creation itself? Is there anything at all physics can say about this? To understand the answer, let's go back to the material of the previous chapter and continue speaking about young Albert Einstein's theory of the universe.

Einstein was a radical. He was not afraid to overturn the most deeply held intuitive beliefs of us common folk. He did this in 1905 with something called the special theory of relativity, which dealt with what happened when things move relative to one another. We will discuss this further in Chapter 23. But it was in 1916 that he took his greatest and most daring step. In this year, he produced the general theory of relativity that dealt with gravity.

Before Einstein, the only theory we had for gravity was that of Isaac Newton who, in 1687, published his *Philosophiae Naturalis Principia Mathematica* (which lazy physicists just call the *Principia*), in which he explained that every object in the universe attracts every other object to it with a force that depends on the masses of the objects and the distance between them. This deceptively simple statement revolutionized physics. It enabled Newton to explain both celestial and earthly phenomena with the same law. It explained the motions of the earth, the moon, and the planets in the same way it explained apples falling from trees. It was an edifice that ruled supreme in physics for over two hundred years, and by the dawn of the twentieth century it could be used to explain everything astronomers had observed except for a very small irregularity in the motion of the planet Mercury that nobody really cared about anyway.

Then, along came Einstein with his pipe and armchair. Einstein knew all about Newton's theory, that all objects attract each other. But, he wondered, *why* do they attract each other? No one had really asked this question before. He also wondered about elevators.

If we get into an elevator, and it suddenly starts to go up, we experience something that feels like a momentary increase in our weight. It is almost as if the force of gravity increased for a short time, but we know that it didn't. What actually happened is similar to what happens if we sit in a sports car and floor the accelerator. The car leaps forward, and we are pressed backward in our seats. The sensation of increased weight in the elevator is caused by the same effect. The elevator accelerates upward, and we are pressed downward toward the floor.

So far we have explained elevators in an obvious way as Newton would have. But Einstein was not satisfied with this. "Look," he said, "suppose there is a physicist standing in the elevator, and we disconnect the little lights that tell him what floor he is on. He experiences a force pressing him toward the floor. Since he doesn't know if the elevator is moving, he cannot tell whether this force is due to the upward acceleration of the floor, or whether there really was an increase in the force of gravity." Newton would have said, "Tough luck." Einstein said, "Since he cannot tell the difference between acceleration and gravity, they must really be the same thing."

This is the essence of the general theory of relativity. According to this theory, gravity and acceleration are really the same thing. The force we seem to feel pulling us downward toward the earth is not a force at all. It is an illusion. It is similar to the mysterious force that seems to push us back into the seat of the accelerating sports car. There is really no force. It is actually the acceleration of the sports car that is doing it. Thus, Newton was wrong. There is no force of gravity; there is only an illusion of force due to acceleration.

This is what we meant when we called Einstein a radical. This idea sounds kind of crazy at first. After all, here we are sitting on the surface of the earth. We are not going anywhere. How could we be accelerating? Well, Einstein was never afraid to take an idea to its logical conclusion. According to his theory, the entire earth's surface with us on it must be accelerating outward. "OK," he said, "we don't think we are going anywhere. That must also be an illusion."

To understand why this is so, we must draw a diagram. Figure 11-1 shows a person who is standing still on the surface of the earth. The vertical direction represents space: ordinary distance in the up–down direction. The horizontal direction represents time, which flows from left to right. Whenever we discuss either of Einstein's theories of relativity, we quickly end up discussing space and time. In our daily lives, we think of space as having three dimensions: length, width, and height. Einstein and a mathematician named Hermann Minkowski showed that it is more correct to think of the universe as having four dimensions, with time being the fourth dimension of *spacetime*. Since we cannot draw the full four-dimensional spacetime on a sheet of paper, we have instead drawn a two-dimensional spacetime with one spatial dimension and one time dimension.

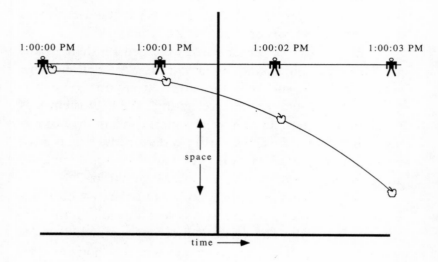

Figure 11-1: The World Lines of a Stationary Person and a Falling Apple

In our diagram a person is standing at a certain place at 1:00:00 PM. One second later, the person is standing at the same place, and so on for 1:00:02 PM and 1:00:03 PM. We have drawn a little picture of this person at each of these times and connected the pictures by a line. This line represents the path the person traces in spacetime just by remaining at a fixed point in space. It is called the *worldline* of the person. Now, suppose the person drops an apple. It starts falling. We have also drawn its worldline.

If Isaac Newton were to view this diagram, he would observe that objects with curved worldlines are accelerating, and those with straight worldlines are not. The man, who is standing still, has a straight worldline because he is not accelerating. He feels the force of gravity, but is not moved by it because the ground is supporting him. To Newton, the apple falls because of the force of gravity. This force makes it fall downward with constantly increasing velocity, that is, constant acceleration. Therefore, when we draw its worldline we find that it is curved.

Now let's see what Einstein has to say about all this. In his theory, there is no such thing as a force of gravity. The reason that the man feels a force (his weight) is because he is accelerating upward. In fact, the entire surface of the earth is accelerating upward and taking him with it. But since there is no force of gravity, the apple, which is following the path of least resistance by falling downward, is *not* accelerating!

Now just a minute here! Can't we see that the worldline of the man is straight, indicating that it is not accelerating, and that of the apple is curved, indicating that it is? What would Einstein say about this? "Simple," answers Albert, "just assume that the entire diagram is printed on a sheet of rubber. Now, stretch the rubber so that the worldline of the apple straightens out and the world line of the person curves. Now he is accelerating and the apple isn't!"

To say something as preposterous as this, a person would have to be either incredibly stupid or an absolute genius. Einstein seems to have been the latter. To Isaac Newton, and to all physicists before the year 1916, space and time were simple, fixed, and rather uninteresting ideas. The spacetime of Isaac Newton was similar to the piece of paper on which our figure is printed: flat and passive. Einstein relaxed the assumption of flat, unchanging spacetime. He realized that, if gravity is actually nothing more than acceleration, the spacetime in the neighborhood of a mass, such as the earth, must be distorted so that worldlines that look straight to us are actually curved and some lines that look curved are actually straight. In Einstein's theory, spacetime itself is active and can be distorted or curved, much as a sheet of rubber.

In the general theory of relativity, space in the neighborhood of bodies with mass, such as the earth or sun, is curved in predictable ways. The circular orbit of a satellite around the earth is actually the shortest distance between any two points on its orbit, even though our intuition tells us that connecting these points with a straight line would give a shorter path. Our intuition is wrong because we tend to think about flat three-dimensional space rather than curved four-dimensional spacetime. It turns out that this distortion of spacetime enabled Einstein to explain the small irregularities observed in the orbit of Mercury. In the neighborhood of extremely massive objects, spacetime can become so distorted that space and time get mixed up, and what we might perceive as time is actually space and vice versa.

These are not easy ideas. It is hard to develop an intuition for what is going on. When physicists have trouble with their intuitions, they retreat into mathematical equations.

Einstein did this in 1916 when he formulated his original theory. Soon afterward, Einstein and the physicist Alexander Friedmann began to apply the mathematics of general relativity to the universe as a whole.

Friedmann asked: What kind of universe is possible according to general relativity? He was able to solve the equations of general relativity and determine that several types of structure are possible for the universe as a whole, but that exactly which one is correct depends upon the amount of mass in the universe—a number he did not know and about which there is much debate today. But there is one quality that all of Friedmann's possible universes have in common. They refuse to stay still. They all display either expansion or contraction with time. A static, unchanging universe is just not possible according to Einstein's original theory.

As we mentioned in the last chapter, Einstein was not happy with this. At the time, most cosmologists supported the steady state theory of the universe. So, Einstein made a change in his theory that would make a steady state universe feasible. A few years later, Hubble showed that the universe is indeed expanding, just as Einstein's original theory predicted. Einstein returned to his original theory, and most physicists now agree. Thus, the expanding universe is explained by Einstein's theory of gravity.

Even though we compared the expansion of the universe to the explosion of a firework, there is a big difference. A firework explodes into the space surrounding it. Space provides the room for the explosion to occur, but it is passive. It does not participate in the explosion. The expansion of the universe, as predicted by the theory of general relativity, is of a much different nature. Einstein and Friedmann were able to show that, in the case of the universe, it is space itself that expands.

Let's return to our analogy of blowing up a balloon. Imagine that initially the balloon is very small. There is a small ant walking on the surface of the balloon. Its world is two-dimensional and curved, much like the surface of the earth. It has finite size but no boundaries. If the ant continues walking in one direction, it will go all around the balloon and eventually return to its starting place. Now let's begin blowing up the balloon. The two-dimensional space that the ant is exploring begins to expand. The expansion takes place everywhere equally. As we saw above, dots painted on the balloon would recede from one another due to the expansion of the two-dimensional space in which they exist.

General relativity tells us that a similar thing is happening to the universe. From the moment of creation, space itself has been expanding. The galaxies that Hubble observed are getting further apart because the space in which they exist is expanding.

If the amount of mass in the universe is large enough, Friedmann's analysis shows something quite amazing: that the universe would have a finite volume, yet would have no boundaries. It would be curved much as the surface of the balloon, so if we were to travel a long distance in any one direction we would go all around the universe and return to our starting point. Such a structure is called a "closed universe." If the mass in the universe is less than this, the volume of the universe would be infinite, and the universe would be "open." In either case, the nature of the the expansion would be the same.

According to current observations, the amount of visible matter in the universe (stars, galaxies, gas clouds, and so on) is not large enough to make it finite. This bothers some physicists, so they are searching for hidden matter, the so-called *dark matter*, that would make the mass large enough to close the universe. This is a major area of astronomical research today.

In the case of a closed universe, the history of the expansion can be represented as the diagram in Figure 11-2. (A somewhat more complicated diagram would be needed for an open universe.) Again, since we cannot draw four-dimensional spacetime, we have instead shown one dimension of space and one of time. Unlike Figure 11-1, space is now curved. The circles represent space, and lines coming from the center of the diagram indicate time. At any given time, all of space is indicated by a single circle a given distance from the center. We have also shown the worldlines of some galaxies and quasars.

Immediately after the big bang, that is, near the center of the diagram, there are many quasars. Sometime later, as we go outward from the center, the quasars (whatever they are) have disappeared and the universe is filled with galaxies. We see that as we go forward in time, away from the center, the circumference of the circles increases, space expands, and the distances between the galaxies increases. Our present time is shown as the outermost circle. There are many galaxies, and they are still moving away from each other.

The very center of the diagram represents the big bang itself. Time proceeds outward in all directions from this point. According to Einstein and Friedmann, not only is the big bang the origin of the universe, but it is also the *origin of time itself*.

This bears repeating. Some people ask, what was going on before the big bang? From the diagram we see that there is no "before the big bang." According to general relativity, time does not extend indefinitely into the past. It originates at the moment of creation. There is no time before this event.

What of the big bang itself? We have already seen that as we get closer and closer to creation, the density and tem-

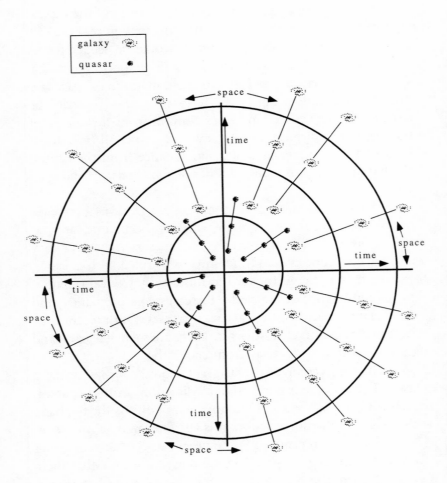

Figure 11-2: The Expansion of the Universe

perature of the universe become unimaginably large. No laws of physics that we now know can explain what was happening. As we have pointed out, this does not mean that there were not some other yet-to-be-discovered laws operating. But general relativity goes one important step further. It tells us that at the moment of the big bang, the density and temperature of the universe were not only very large, they were infinite. The moment of the big bang is what mathematicians call a *singularity*,[5] a point at which mathematical infinities occur. But, true infinities cannot really occur in the physical world. Something is wrong here. Therefore general relativity itself breaks down at the beginning of the universe. The big bang cannot be explained by the laws of nature.

The creation of the universe was not a reforming of that which was already in existence, a *yatzar*. Indeed, there was no existence before the big bang. The creation of the universe was a unique event that cannot be explained by the laws of physics. It was the origin of time itself. It was a true *bara*. We see a striking parallel between what Torah and modern physics have to say about the origin of the universe.

Let's return to the blackboard on which we wrote the events of creation according to the Torah. We are going to write a second list on the board, one that summarizes the events of creation according to modern science. The first entry of the Torah side said, "Moment of creation, not describable by the laws of physics." To start our second list, we write:

---

[5]It is true that some physicists, such as Stephen Hawking are trying to find a way around the singularity, but their efforts are based upon combining general relativity with Quantum Mechanics and are still highly speculative.

1. Big bang

which matches exactly. What better way is there to summarize the situation but to say: In the beginning *Elokim* created
(*bara*) the heavens and the earth."

12

Very Small Things
Quantum Theory

Our grandmothers did not need fever thermometers. When their babies looked a little listless, they pressed their lips to little foreheads and unerringly determined whether or not the babies were ill. But we are more comfortable with little glass tubes filled with mercury. How do they work?

When we place a thermometer in our child's mouth, the mercury within it warms up to her body temperature and expands. The warmer it becomes, the more it expands. The expansion causes it to move up a small channel, and this enables us to read the temperature.

In the last chapter, we discussed the laws of thermodynamics. What do they tell us about all this? When we take the child's temperature, the mercury in the thermometer becomes warmer by gaining heat energy. The first law of thermodynamics tells us that this heat must come from somewhere, and that place is the child's mouth. Some of the heat energy in the child's mouth is transferred into the thermometer. As a consequence of this, her body loses a small amount of heat and this causes the little girl's body to become a tiny bit cooler. But the child's body is so much larger than the thermometer that the amount of heat energy she loses is very small in comparison to all the heat in her body. Thus, the change in her temperature is minuscule, less than a thousandth of a degree. We don't care about it, and we never notice it.

Suppose, instead of measuring the temperature of a child, we are measuring the temperature of a teaspoon of water. Again, the thermometer becomes warmer, and the water becomes cooler. But now, the total amount of heat in the water is not very large compared to that of the thermometer. The temperature change of the water may not be large enough to be noticeable. In the end, we determine not what the original temperature of the water was, but rather what the temperature of the water became *after it cooled* as a consequence of giving up some heat to the thermometer.

Finally, suppose we are trying to measure the temperature of an ant. The thermometer has much more heat capacity than the ant. If we bring them into contact (letting the ant walk around on the tip of the thermometer is probably the best we can do), the ant undergoes a large temperature change as it gives up heat to the thermometer, but the temperature of the mercury changes very little. We get almost no information about what the ant's temperature was initially.

We have seen two things. First, that the act of measuring the temperature of something changes the very temperature we are measuring, and second, that the smaller the object is, the more its temperature is affected by this. These effects are not unique to thermometers or thermal measurements.

For example, suppose we wish to determine where something is. One way we can tell is by looking at it. We cannot see it in the dark, so we must allow light to fall on it. We know that light consists of little particles called photons. (Let's forget about the wave theory of light for a moment.) When these photons bounce off the object, they jostle it around a little bit. Therefore, to measure the position of a particle, we necessarily disturb it in some way. The smaller the object is, the more it is disturbed by the act of measurement.

In all cases, the act of measuring a physical quantity causes a disturbance, and the smaller the object is, the greater the disturbance. This may seem obvious now that we have thought about it, but in 1927 it was a new idea. A 26-year-old physicist named Werner Heisenberg first noticed this and stated it as a physical law.

To understand Heisenberg's formulation of this law, let's again consider the act of measuring the position of a particle, such as an electron. Suppose the particle is moving. We shine light on it so we can see it. This is equivalent to bombarding it with photons. As the photons bounce off, they transfer small amounts of energy to it and change its velocity. At any given time, we can determine where the particle is, but there is a cost: in so doing we will necessarily change its velocity in some random way. Therefore, although we can determine where it is now, we will not be able to predict exactly where it will be in the future.

To measure its position with some degree of precision, we necessarily disturb its velocity. Heisenberg was able to show that we could also set up an experiment to measure its velocity precisely, but that in so doing we would necessarily change its position in some random way. In other words, we can determine precisely *either* the position of a particle or its velocity, but not *both*.

This law of nature is called the *Heisenberg uncertainty principle*. It is one of the foundations of *quantum theory*. Heisenberg observed that measurable physical quantities come in pairs. Velocity and position are one such pair. The more precisely we measure one member of a pair, the more uncertainty occurs in the other. The effect becomes most noticeable when the objects involved are small. If we measure the position of an elephant, we do not make a very large disturbance in its velocity. But if we do the same to an atom, we may disturb its velocity so much that it goes zipping off in a totally unpredictable direction. Quantum theory is not really important when it comes to understanding the fairly large objects we deal with in our everyday lives, but it is vital for understanding the behavior of very small things.

As a consequence of Heisenberg's principle, we are always uncertain about what is going on in any physical system. We can never learn enough about it to completely predict its future behavior. We can say what it will *probably* do, but if it is composed of microscopic particles, we can never be sure of what it will *certainly* do. We will always be unsure of the future behavior of very small things.

There is a subtle question here. Just because *we* can't predict what will happen, does this mean that *nature* itself is really unpredictable? Could it be that there is really something predictable going on, but that it is our inability to make precise observations that makes it impossible for us to predict what will happen?

To some extent, this question does not matter. Science can only deal with that which is measurable and observable, anything else is outside the domain of science. Therefore if *we* cannot predict the future of the world, we might as well regard it as unpredictable. In the words of the philosopher

Ludwig Wittgenstein, "that of which we cannot speak, we must pass over in silence."[1]

Nevertheless, most of us are driven to wonder what is really going on, and so we ask: "Does the world really act in a random fashion?" The idea that the universe might be intrinsically unpredictable was troublesome to physicists. From the time of Isaac Newton, physicists had believed that, armed with the laws of physics, they could take any physical system and predict its future behavior. This was called the era of classical physics. The practice of classical physics over a two-hundred-year period led to spectacular successes in understanding the motions of the planets, the nature of electricity and magnetism, and the behavior of heat. But in 1927 it all came crashing down. Physicists were forced to come to terms with an intrinsic limitation in their ability to predict nature.

Many did not take it lying down. Indeed, Einstein said that he could not believe that "God plays at dice." He spent the majority of his life trying to find a way around the quantum theory. He was unsuccessful. In 1964 a physicist named John Bell proved a theorem that cast a pall over efforts to save classical physics. Bell's theorem enabled experiments to be done that confirmed the unpredictability of nature at its deepest level. It is not just our our inability to make measurements that makes the world unpredictable, it is the nature of the universe itself.

The implications of this are astounding. In the first place, let's consider the nature of a small object, such as an electron. If we measure the electron's velocity precisely, we can have no knowledge of its position. Since the uncertainty principle

---

[1]Quoted in J. E. Dodd, *The Ideas of Particle Physics* (Cambridge: Cambridge University Press, 1984), p. 19.

is not a statement only about our inability to know reality, but about reality itself, this means that the electron actually *does not have a position!* But doesn't a particle have to be *somewhere?* Yes, it does. So in this case, the electron is not acting like a particle. Since it does not have a definite position, it is somehow spread out in space. The electron seems much more similar to a wave than a particle. Indeed, an electron whose velocity has just been measured can be shown theoretically to behave as a wave moving through space. On the other hand, if we had chosen to measure the position of this very same electron instead of its velocity, we would have gotten an answer indicating that it *was* at a certain point in space. In this case, the electron would behave as a particle rather than a wave.

Electrons and all other small things behave as either particles or waves. It all depends upon which experiment we choose to do. What, then, is an electron *really?* In the world of quantum physics, this question has no easy answer. An electron is, in essence, an electron. More than that we cannot say. It is intrinsically neither a particle nor a wave. It is only when we observe it that it temporarily becomes one or the other. The same is true of all other bits of matter, and it is also true of light, a fact we have already mentioned in Chapter 5.

To secular physicists, who see nature as the ultimate reality, this is very disturbing. But from the viewpoint of Torah, there is no problem at all. Let us recall the concept of the *tzimtzum*, that God conceals himself in the world. Why does an apple fall from a tree? It falls because God makes it fall. But since God causes the apple to act in a manner that we can predict, we can choose to ignore God completely and say that it is only a law of nature that makes the apple fall.

Quantum theory does not disturb us. An electron has no reality other than that which God wills. If we measure its position, He arranges for us to get an answer consistent with it being a particle. If we measure its velocity, we find it to be a wave. As we say in the moving poem on the evening of *Yom Kippur*, "like the glass in the hand of the blower, He shapes it at will and dissolves it at will."[2]

We can still choose to ignore God if we wish. There are still physical laws. It's just that they are not as satisfying as we would like them to be. The universe acts in a manner that we cannot predict. We can choose to realize that God is making things happen, or we can choose to say that nature is random. It is all up to us.

At one time, scientists thought that quantum theory implied that nature was unpredictable only in the microscopic world. In the everyday realm of large objects, quantum uncertainties were assumed to be so small that we would never notice them. This would explain why our intuition rebels when we encounter these phenomena—we have been trained from birth in the world of large things. During our formative years, everything looked deterministic. Of course, there were always a few things we couldn't predict in detail, such as the weather or the behavior of our little sister. But we always had confidence that someday scientists would be able to forecast the weather accurately, and as for our sister, well, the less said about her the better.

For a long time, physicists thought that this was correct, that quantum effects were too small to affect the everyday world. The reason for this is that most physical systems are so stable that they are insensitive to small disturbances. Consider, for example, a quiet lake. We throw a tiny pebble into

---

[2]Rabbi N. Scherman, *The Complete Artscroll Machzor—Yom Kippur (Nusach Sefard)* (New York: Mesorah Publications Ltd., 1986), p. 133.

this lake. Ripples spread out in all directions until the entire surface is in motion. But the disturbance is small, and its effect soon disappears. The oscillations die down and the lake is again quiet. The small disturbance has had no lasting effect on the lake, and in a short time all memory of it is gone.

But there are some cases in which small disturbances can have lasting effects. Let's consider a marksman who is aiming a rifle at a target. He is about to shoot. If, as he squeezes the trigger, we gently nudge the side of the rifle barrel, we can cause him to miss the target. The effect of our small disturbance is magnified. Yet, the amount of his miss is proportional to the magnitude of our disturbance. The larger our nudge, the larger his miss. Conversely, if we make an extremely small disturbance, it will cause only a minor perturbation in his aim. A truly microscopic quantum uncertainty would have no noticeable effect.

Are there any systems in which even a microscopic disturbance will have a major effect? Yes. Let's consider a knife balanced on its point. It is in a very precarious position. A small puff of breath will send it crashing to the table. Such a system is called unstable. No matter how small the disturbance, the knife will still fall over. If the knife is very sharp, and there are no external disturbances, the quantum mechanical uncertainty in its velocity due to our act of observing it will be enough to make it fall.

Recently, a new field of mathematics called the *theory of chaos* has shown that many large physical systems have similar instabilities in their behavior. They are as sensitive to small disturbances as a knife balanced on its edge. But the effect of small disturbances on such chaotic systems is not always as obvious as is the case of the knife.

Let's consider the weather, for example. Many believe it to be a chaotic system. This means that it is so sensitive to small disturbances that, although we can do a pretty good job of predicting the weather over the short run, we cannot do so over longer periods of time. The very small change in the velocity of the wind caused, for example, by a person raising his hand in a windstorm, could eventually change the behavior of the weather in a major way. For the first few days, the effect would not be great, but in the long run the effect might grow and could ultimately mean the difference between rain and shine. It is for this reason that it is unlikely that we will ever be able to predict a year in advance whether or not it will rain on the day of the sisterhood picnic. In the words of one chaos theorist: "a butterfly beating its wings in a Brazilian rain forest could influence the weather over the northern hemisphere for the next century."

A rabbi in Baltimore once joked that he experienced a crisis in his belief from driving through the state of Oregon. "The Torah tells us," he said, "that everything in the world, every insect, every leaf, has a purpose. As I drove through the miles and miles of forests and saw the billions of leaves, I wondered: how could each one be important?" Chaos theory tells us how. The presence or absence of a single leaf in a forest in Oregon could trigger a flood or a drought in an Asian country a hundred years later.

Chaotic systems are so sensitive that even a small fluctuation in the wind velocity due to the quantum theory's uncertainty principle could mean the difference between sunny skies and a hurricane after ten years. Maybe those television weathermen really are doing the best they can.

So much for the weather, but what about our little sister? Can quantum uncertainties affect her behavior?

From the viewpoint of classical physics, a person's brain is a machine just as a computer is a machine. (Of course, electronic computers did not exist during the classical era of physics, the years before 1927, but we have taken the liberty of phrasing our discussion in modern terms.) A few decades ago, Norbert Weiner coined the term *cybernetics* to describe the study of the similarities between computers and the brain. Now if a brain is a machine, then, according to classical physics, we can eventually discover the laws governing its components. Then, if we are given complete information about its state at a given time, we should be able to predict its behavior in the future. Of course, this is a formidable task, but it would be possible in principle. This caused a problem for those philosophers who believed in free will. If a person's future behavior is predictable, then where is the room for free choice?

This philosophical problem no longer exists in modern physics. The brain is a complex device. (In fact, it has been argued that the human brain is *the* most complex device known.) It is quite likely that brains are chaotic systems. If so, a small quantum uncertainty in the electrical charge on a cell in the brain of a guinea pig could cause a difference in its ultimate choice of a mate. Small random quantum fluctuations could make it impossible for us to ever predict human behavior with any accuracy. Maybe this is why we have never understood the behavior of our sister.

Why, from a Torah viewpoint, do quantum fluctuations in inanimate objects occur? They occur because God wills them to occur. Just because we cannot predict God's actions, it does not mean that they are random or without purpose. No, Dr. Einstein, the quantum theory does not mean that God plays at dice. It just means that we cannot predict God's actions.

What of our free will? The Torah tells us that God has granted us absolute freedom of choice. The quantum fluctuations in our brain cannot be predicted by another person, but that does not mean that they are random and meaningless. What the other person may be unable to predict might be our exercise of free will. The psalm says, "The heavens are *HaShem's*, but the earth He has given to man."[3] Perhaps we should add: "all quantum fluctuations are *HaShem's* to control, but those in our minds He has handed over to us."

We have found the behavior of very small things to be unpredictable, but this does not disturb us. It just means that the world is a very different place than some scientists would like. In the next chapter we will discuss the behavior of very fast things and find that the world is stranger still. Then, in the following chapter, we will talk about things that are both small and fast, and we will return to the story of creation.

---

[3]Psalms 115:16.

13

Very Fast Things
The Equivalence of Mass and Energy

Einstein sat around a lot. It wasn't that he was lazy; it's just that he was a theoretical physicist rather than an experimental one. Instead of conducting experiments in a laboratory, he would conduct them in his mind. He called these *gedanken experiments*, which means thought experiments in German. He would start from a few simple assumptions and then imagine what would happen if a physicist tried to conduct a particular type of experiment under these conditions. He would analyze the situation and try to predict what the physicist would observe. Based upon these *gedanken experiments*, Einstein constructed his two theories of relativity, which have now been tested and verified through real laboratory experiments.

In Chapter 11 we discussed the general theory that he published in 1916; in Chapter 23 we will discuss the special theory that he formulated in 1905. (Since the Torah is written in conceptual rather than chronological order, we have taken the liberty of doing the same thing in this book). Following his publication of the special theory of relativity, Einstein published a short paper with the catchy title: *"Ist die Traegheit eines Koerpers von seinem Energiegehalt abhaendig?"* [1] ("Is the inertia of a body dependent upon its energy content?") It is in this paper that he first derived his most famous formula.

Before the work of Einstein, physicists had well-defined ideas about mass and energy. They were regarded as very different things.

Mass is a property of all matter. On the surface of the earth, the mass of an object is roughly equivalent to its weight, so heavy objects have larger masses than lighter ones. But mass is a property that objects have even in a weightless environment. Technically speaking, the mass of an object tells us how much inertia it has, that is, how hard it is to start it moving and, once it is in motion, how hard it is to stop it. If an astronaut could take a grand piano with him in the space shuttle, he would find it to be weightless. But a piano is an object with a large mass. The astronaut would have to give it a formidable shove to get it moving in the direction he wished, and another astronaut could be injured by it if he got in the way of it once it was moving.

Energy is a little harder to understand. Energy is a quantity associated with the motion or configuration of objects. Moving bodies have energy associated with their motion. This is called kinetic energy. Compressed springs can potentially cause other bodies to move once the springs are released. Therefore, we say that they have potential energy. There are

---

1. A. Einstein, "Does the Inertia of a Body Depend Upon its Energy Content," in *The Principle of Relativity*, trans. W. Perrett and G. B. Jeffery (n.p.: Dover Publications, n.d.), pp. 67–71.

many other forms of potential energy: chemical energy, nuclear energy, and the energy of a roller coaster slowly rounding the highest point on its track and ready to drop. We have seen that heat is also a form of energy.

For the two hundred years preceding the twentieth century, physicists regarded energy and mass as independent quantities. They spoke of two conservation laws. The conservation of mass meant that mass could be neither created nor destroyed. The total mass of a physical system was assumed to be fixed, and nothing could ever change it. Similarly, energy was conserved. The total amount of heat, potential energy, and kinetic energy was constant. Kinetic energy could change into potential energy, potential into kinetic, and they could both turn into heat, but the total energy could neither increase nor decrease. Everything was very simple.

Then, along came Einstein and his thought experiments. He thought very carefully about what would happen when objects moved very fast, nearly as fast as light. He was able, through these thought experiments (which we cannot explain here), to show something quite shocking: that mass and energy are equivalent. We all know the formula $E=mc^2$. It means that an amount of energy (E) is equivalent to an amount of mass (m) multiplied by the square of the speed of light (c). Mass can be converted into energy and vice versa. How could this be? Intuitively, mass is associated with the material existence on an object while energy is a mathematical curiosity that tell us something about what the object is doing. How can one turn into another? Einstein proved that our intuition is wrong.

Since the speed of light is a very large number, this formula implies that a small amount of mass can be converted into a huge amount of energy. This is the basis of nuclear power—and nuclear weapons. It was anticipated by Einstein in his original paper on the subject: "It is not impossible that with bodies whose energy-content is variable to a high degree (e.g., with radium salts) the theory may be successfully put to the test."[2]

A small amount of the mass in the nucleus of an atom can be converted into energy, and this energy can be converted into electrical power—or it can be used for great destruction. The choice is ours. Not only has modern physics changed our view of the universe on the theoretical level, but it has also put formidable powers of creation and destruction in our hands. Science provides the *is*, but not the *ought*. It gives us the power, but it doesn't tell us how to use it. At one time, we had the luxury of not caring about this, but the old rules of human history are changing. We desperately need a user's guide. Fortunately, we have one, as we pointed out in Chapter 4.

Back to physics. A moving body has kinetic energy. Einstein's formula tells us that this kinetic energy is equivalent to added mass. Therefore the mass of the body when moving is larger than when it was still. In colloquial terms: objects become heavier when they move. In normal life, this effect is so small that we never notice it. A 200,000 pound jet airplane moving at 600 miles per hour has its weight increased by only a millionth of an ounce. Yet when bodies move very fast compared to the speed of light, their masses increase dramatically. Since the mass of an object tells us how hard it is to change its speed, objects moving faster and faster require larger and larger forces to keep them accelerating. As a body approaches the speed of light, its mass increases without limit. If it could reach the speed of light, its mass would

2. Ibid, p. 71.

become infinite, but since it would take an infinite force to make this happen, such a thing is impossible. Therefore, no object can ever be accelerated to the speed of light. There is a fundamental speed limit in the universe. Our intuition may rebel against this, but modern physics tells us that it is so.

Very small things, very fast things. The study of both has radically changed the way that physicists view the world. But so far we have been discussing quantum theory and special relativity separately. It is when we put them together that really interesting things happen. This is the subject of the next chapter.

14

Nothing
The Chaotic Nothingness Underlying the Universe

The largest pieces of laboratory equipment in the world are called particle accelerators. Some of these devices are thousands of feet across. In them, small subatomic particles are accelerated to velocities very close to the speed of light. They then have very high kinetic energies and correspondingly large masses. These particles are aimed at each other, and physicists carefully observe what happens when they collide. In this way, they learn deep lessons about the basic forces that operate in the universe.

A favorite experiment involves shooting energetic electrons at the protons in the centers of hydrogen atoms. By observing the energies and directions of electrons scattering off the protons, physicists can learn about the nature of the forces between them.

The experiments seem quite simple. There are only two players in a collision: an electron and a proton. It is very easy for physicists to predict what will happen when one particle bounces off another. Yet, when they sharpen their pencils,

161

perform their calculations, and compare their predictions to the actual experimental results, they find something quite disturbing. The electrons do not act as if they are bouncing off single protons. Instead, they act as if they are bouncing off a whole group of small particles in addition to the protons. During the short time of the collision, there seem to be other particles hanging around. Where do *they* come from? Where do they *go* afterwards?

Subatomic particles are very small things. They are also very fast things. To understand what is going on, we have to combine quantum theory and relativity.

We start by considering the Heisenberg uncertainty principle. Physical quantities come in pairs, such as velocity and position. If we determine one member of a pair accurately, the other ceases to have a definite value. Once we know the velocity of a particle accurately, the particle no longer has a definite position. It turns out that energy and time also form such a pair. If we measure the energy involved in some event very accurately, we cannot be sure of exactly when this event occurred. Conversely, if we make a very precise measurement of when something happened, we cannot be sure of how much energy was involved.

Electrons move very fast. When an electron collides with a proton, it is close to it for only a very short period of time, typically less than 10^{-23} seconds. Therefore, when we observe the outcome of an electron-proton collision we can have very accurate knowledge of when this encounter occurred. Since we know the time of the encounter very precisely, the uncertainty principle then tells us that we cannot be very sure of how much energy was involved. There could be a little extra energy hanging around, energy that sort of comes from nowhere.

What about the conservation of energy? Isn't it true that energy cannot spontaneously appear or disappear? Yes, this is true, but only in the long run. If we observe the electron and the proton over an extended period of time, we will see that their total energy does not change. There is no extra energy available. But when we look at an electron and a proton for a very short time, such as during the brief encounter of a collision, we enter the Alice in Wonderland realm of quantum mechanics, and things become very unpredictable. Their total energy is no longer certain. There may be a little extra around. Physicists say that they borrow some energy from . . . from where? It doesn't really matter where they borrow it from, since this additional energy doesn't really exist in any definite sense! (This really does sound like Alice in Wonderland, doesn't it?) Yet, for the short time they are together, there seems to be a little extra energy around. It's just that if we continue to look for a while, this energy disappears since it was never there in the first place.

This all sounds sort of silly, almost as if we are joking. But we are not. This is an accurate account of the sort of discussions that take place in the hallowed halls of universities teaching modern physics. Nobody is really happy with this situation. Most physicists would like to have a much clearer understanding of what is going on, but it seems that God has just refused to make the universe behave in a manner fully understandable by man.

So, during the short period the electron and the proton are close together, there is some extra energy present. But from the special theory of relativity we know that energy and mass are equivalent. Therefore, some of this extra energy can turn into matter—some extra particles can spontaneously appear from nowhere. Physicists call these *virtual particles*, meaning that they don't really exist. They will later disappear, as they must since they cannot have been there in the first place, but for the little while they are around, they can cause a bit of mischief. In addition to colliding with the proton, the electron can also collide with these virtual particles. This is why electrons act as if they are bouncing off a whole bunch of particles even though they are only colliding with single protons.

What happens is the following: A single electron approaches a single proton. They are close together for only a very short time. During this short time, a whole bunch of virtual particles spontaneously appear from nowhere. The electron then interacts not only with the original proton but also with this cloud of virtual particles. Because of these virtual particles, the electron bounces off in a different direction than it would have if only a single proton were present (what physicists would call a naked proton, one without its cloud of virtual particles). After the collision, the virtual particles disappear. Well, they don't really disappear; it's just that they were never there in the first place.

We must stress that this is not idle speculation. Physicists have performed real experiments with electrons and protons. The electrons do *not* behave as if they are bouncing off simple naked protons. There are other things around. Physicists have also performed many other experiments with different particles. The conclusions have always been the same. We now have hard experimental evidence of the existence of virtual particles that do not exist! Who says that God does not have a sense of humor?

Physicists building the next generation of particle accelerators hope to see entirely new types of particles that do not exist. But we do not need such an expensive machine to learn something very important. Like Einstein, we can sit back in our armchairs and perform a *gedanken experiment*; we can do an experiment in our minds.

Instead of thinking about a collision between two isolated particles, let's consider a physical system that is even simpler. Actually, it is the simplest physical system we can conceive of: nothing! Suppose we have some empty space. Nothing is there, but the laws of physics are in effect. If something were there, it would be subject to the laws of nature. It's just that nothing is there in the first place. Physicists call this the *physical vacuum*.

What do we see if we observe the physical vacuum very carefully over a period of several days? Nothing. Why? Because nothing is there. So far, so good. But now, let's consider what happens during a very short period of time. Because the time is very short, there is an uncertainty in the energy. The additional energy that may be present can turn into matter, and a cloud of virtual particles can appear. What happens is exactly the same as what occurred during the collision of an electron and a proton; it's just that in our present case there are no real particles around.

During any short period of time, the physical vacuum is filled with virtual particles. What particles are present? Any and all. During increasingly short intervals of time, there are greater and greater uncertainties in the energy of the vacuum. Therefore, larger and more massive particles can appear, just as long as they disappear soon afterwards. Nothing is there, yet everything is there. Modern physics is almost beginning to sound like mysticism.

What does all this have to do with creation? Quite a bit. For one thing, the paradoxical nature of the physical vacuum is very similar to a certain paradox we encountered in our study of Genesis. We will discuss this in greater detail at the end of this chapter (no fair looking ahead). But before we get into this, let us recall that the Torah deals with two different aspects of creation. First, there is the original moment of creation, the big bang, the origin of time and space; and second, there is also the quieter constant renewal of creation by God at all times. We will see that our discussion above relates very closely to both of these aspects. Let us begin by considering the big bang.

We have seen that creation cannot be described by the laws of physics. But the laws of physics *are* applicable at any time following the big bang. By taking the current state of the universe and extrapolating backward, it is possible to say some intelligent things about what was happening. As we have seen in Chapter 11, most cosmogonists feel that we are in pretty good shape when it comes to understanding what happened a few seconds after the big bang. At this time all the matter and energy in the universe were compressed together at incredibly high density and pressure.

We are familiar with the concept of pressure from everyday life. We determine when our car tires need air by using a pressure gauge. If the pressure is too low, the tire is flat. We correct the situation by connecting a hose and pumping in air. This makes the tire bulge outward and helps support the weight of the car. But we have to be careful. If we pump in too much air, the pressure will increase to a dangerous level, and the tire will burst.

Pressure makes things burst. It pushes things outward. It is the pressure of hot gasses that makes a bomb explode. In the first seconds following the largest explosion ever, the big bang, the pressure in the universe was enormous. We know that at this time the cosmos was beginning the rapid expansion that continues to this day. It would be entirely reasonable for us to assume that the immense pressure contributed to this expansion, that indeed the enormous pressure might be what caused the universe to fly apart.

It would be reasonable for us to assume this, but unfortunately we would be wrong! The conditions of the early universe were just too different from those of our everyday experience for our intuition to be of any use. We must instead rely on the principles of theoretical physics. Let's think about extremely high pressures in the context of relativity theory.

Suppose we start pumping air into a tire, a really strong one that will not burst. It takes energy to push the air into the tire. This energy does not disappear; it ends up being stored in the high-pressure air as potential energy. Of course, some of it goes into waste heat—we can never get away from the second law of thermodynamics—but most of it remains in the air.

The high-pressure air has a high energy content. Since we know from the theory of relativity that energy and mass are equivalent, the high pressure must cause the air to increase in mass.

In Newton's theory of gravity, all bodies attract one another. The more massive the bodies are, the stronger their attractive forces. The increased mass of the high-pressure air in the tire causes a small increase in the gravitational force due to air in the tire (which is pretty small to begin with, although we can measure it with very sensitive instruments). But Newton's theory is only accurate when it comes to fairly low pressures and weak gravitational fields. When the pressure gets extremely large, we must use Einstein's theory of gravity, the general theory of relativity.

Now things get a bit complicated. In physics, pressure is an example of a more general quantity called *stress*; and in general relativity it is not only mass that causes gravity, but also stress. It would take several pages of equations to explain why this is so, so we will just have to accept it as a given fact. According to Einstein's theory of gravitation, not only does the added energy in the high-pressure air increase its gravitational force, but the high pressure itself is a stress that acts as a source of increased gravity.

For an automobile tire, these effects are pretty small, but in the early universe the pressure was so high that it acted as a major source of gravitational attraction. It created forces that tended to hold the universe together. Paradoxically, the immense pressure did not contribute to the expansion of the universe. Instead, it slowed it down!

This does not create any theoretical problems during the first few seconds following the big bang—even though the universe was being slowed down it still had enough momentum to continue the expansion—but if we try to go back really far in the history of the universe, to the first 10^{-35} second, we get into trouble. The pressure would have been so high that it would have acted as too strong a brake on the expansion of the universe. It would have stopped the big explosion from continuing.

Until recently, physicists were up against a brick wall when they tried to get around this problem. But, in recent years, an ingenious theory called *cosmic inflation* has enabled cosmologists to explain what might have been going on. The theory of cosmic inflation began in 1980 with the work of Alan Guth.[1] His original theory has since been modified, and

---

[1] E. Tryon, "Cosmic Inflation," *Encyclopedia of Physical Science and Technology*, ed. Robert A. Myers (Orlando, FL: Academic Press, 1987), 3:709–743.

it may be further refined in the years ahead, but the main features of it are likely to remain.

Guth was every bit as gutsy as Einstein. OK, he observed, the high pressure of the early universe acted as a brake on the expansion of the universe. Is there anything that could have acted as an accelerator? Suppose at some still earlier stage the universe did not have a very high pressure, but instead it had a *negative* pressure. Wouldn't this have the exactly opposite effect of a large *positive* pressure? Wouldn't it accelerate the expansion of the universe rather than slow it down?

What do we mean by negative pressure? Positive pressure pushes things outward; negative pressure would suck things inward. This is quite a bit different from what happens when we suck our cheeks inward. In this latter case, we are merely removing air from our mouths and allowing the outside air pressure to push on our cheeks. A true negative pressure would be an actual inward force, something that does not occur in nature in the present universe.

A negative pressure would constitute a negative stress, and Einstein's theory of gravitation tells us that such a stress would cause a gravitational force that is repulsive rather than attractive. Again, the same paradox: just as a positive pressure presses outward but causes a gravitational force that keeps the universe together, a negative pressure would suck inward yet cause a gravitational force that would cause the universe to fly apart. If there were a highly negative pressure in the first small fraction of a second following the big bang, it could have acted as the explosive force that started the expansion.

But negative pressures and repulsive gravitational forces have never been observed in nature. In addition, if all the matter in the universe were compressed to enormously high density, the pressure would have been very high and positive. How, then, could the negative pressure have existed? It is here that Guth's theory becomes truly ingenious. Not only does it explain the expansion of the universe, but also the origin of all the matter and energy within it.

Einstein was not afraid of stretching space a little bit—Guth was not afraid of stretching his imagination quite a lot. If space filled with matter has positive pressure, and empty space, "nothing," has zero pressure, what could have negative pressure? How about space filled with less than nothing?

How could there be less than nothing? Is there any way this could make sense? Yes, there is. We have seen that, in modern physics, "nothing" is very complicated, it is a sea of virtual nonexistent particles. The physical vacuum contains nothing, yet it contains everything. All possible particles are present, at least for a short time. What if we had a vacuum in which some of the virtual particles were missing. Wouldn't this be "less than nothing"?

One of the particles present in today's universe is called the Higgs boson, named after the Scottish physicist Peter Higgs. No one has ever seen a Higgs boson, but there are deep theoretical reasons for assuming that they exist. They turn out to be very special particles that are deeply involved in determining the fundamental forces of nature. Like all particles, Higgs bosons also act as waves. Physicists prefer to refer to them as *fields*, meaning things that spread out in space. We will speak more about these unusual fields in the next chapter.

A vacuum (that is, nothing) has within it, among other nonexistent particles, virtual Higgs bosons. Guth considered what a vacuum would be like without any Higgs bosons. It turns out to be a form of "less than nothing" that has some very interesting properties. For example, unlike the true physical vacuum that has no energy in it, a vacuum without Higgs bosons has a very high energy content, even though there is less than nothing there! Not only that, but it also turns out to have negative pressure. (Just what we have been looking for!) Guth referred to this enigmatic state of nothingness as the *false vacuum*. This may seem somewhat unbelievable, but in the words of physicist Edward Tryon: "Of all the fields known or contemplated by particle physicists, only Higgs fields have positive energy when the fields themselves vanish. This feature of Higgs fields runs strongly against intuition, but the empirical successes . . . force one to take the possibility seriously."[2]

We never see any false vacuum around nowadays, and for good reason. It turns out that it can only exist when the temperature is very high, in excess of 10^{27} degrees. If the temperature is any lower, the false vacuum tends to freeze into the ordinary physical vacuum of our present world. But in the first instants of the early universe, the temperature was indeed very high. It was high enough for the false vacuum to exist.

Not only is the false vacuum a strange state of less than nothingness, but it also turns out to be very unstable, much like a knife balanced on its point. The smallest disturbance will tip it over and cause it to "fall" into a real physical vacuum.

Let's put it all together. According to Guth's theory of cosmic inflation, the early universe the barest fraction of a second after creation (before 10^{-35} second) consisted of a false vacuum. Nothing was present, not even Higgs bosons. Yet everything else was there all mixed up in a chaotic fashion. The energy content of the false vacuum was very high, equivalent to a mass of 10^{72} pounds per cubic inch,[3] the temperature was very high, over 10^{27} degrees, and the pressure was

---

[2]Tryon, p. 732.
[3]Tryon, p. 732.

negative. This negative pressure, sucking inward, produced a repulsive gravitational field, pushing outward, that caused the universe to explode. This was the origin of the expansion that has continued to this day. As the universe expanded, it cooled. Eventually, the cooling of the universe and the instability of the false vacuum, which was disturbed by a small quantum fluctuation, caused it to freeze into the present physical vacuum, our ordinary garden variety of nothingness. Where did all the energy of the false vacuum go, since the physical vacuum has no energy? It went into the creation of all the matter in the universe.[4]

Fantasy? Poetry? No. Modern physics. Where else but in a modern science textbook could a respected scientist such as Edward Tryon say: "quantum uncertainties suggest the instability of nothingness."[5]

Now let us compare what we have learned with the story in Genesis. What does the Torah tell us of the early universe? Let's look at the blackboard. The second verse of Genesis tells us that, following the initial creation, the world was in a state called *Tohu VaVohu*, which we have translated as chaotic nothingness. When we first wrote this, it seemed like a paradoxical combination. How could nothingness be chaotic? But paradoxical as this seems, it is no stranger than what physicists are saying about the early universe.

Physicists have been forced into considering the nature of nothingness, and they have found it to be very complex. The physical vacuum has nothing in it, yet everything is there. The false vacuum, that state of the universe that immediately followed the creation, had even less. It was a true nothingness

---

[4]Tryon, p. 739.
[5]Tryon, p. 743.

beyond our ability to conceive of nothingness. Yet everything was in it, mixed up, chaotic, waiting to be born.

Two paradoxes: the *Tohu VaVohu* of the Torah, and the chaotic false vacuum of the physicists. We may not really understand either, but they are strikingly similar! What clearer characterization of *Tohu VaVohu* could we ask for? On our blackboard, next to *Tohu VaVohu* we write "the false vacuum."

So much for the big bang and its sequel. But there is a second aspect of creation, God's continual renewal of creation at all times. Does modern physics have anything to say about this?

We have seen that the physical vacuum is a chaotic place, with all sorts of particles being constantly created and destroyed. But what of the particles that really exist? What about the protons, neutrons, and electrons that make up all atoms? Aren't they, at least, permanent inhabitants of the universe?

No. The modern theory of what happens when particles interact with each other is called *quantum field theory*. This is a very complex and difficult subject, and we cannot really give an explanation of what it is all about without getting very mathematical, but there is one consequence of this theory that is easy to understand.

Let's consider what happens when a moving electron passes nearby a stationary proton. The proton has a positive charge, and the electron has a negative charge. Opposite charges attract one another so, instead of continuing in a straight line, the electron is diverted towards the proton. (Of course, the proton also is affected by the electron, but since the proton is much more massive than the electron, this is a small effect that we will ignore for simplicity.) If the electron is moving slowly enough, and is close enough to the proton, the attraction of the proton will cause the electron to circle around it, much as the earth's gravity causes a satellite to remain in its orbit. A lone proton being orbited by a single electron is called a hydrogen atom.

So far, we have been speaking intuitively, using the language of classical physics. A single electron has had its direction changed by the force due to the proton. But quantum field theory gives a very different picture of what is happening. It tells us that the incoming electron did not change its direction, since electrons *never* change direction.

Why, then, do we think the electron changed its direction? Because the force of the proton does a very strange thing. Instead of causing the electron to change its direction, it actually destroys the original electron and causes a new one to appear that is moving in a different direction. The destruction of the original electron and the creation of the new one occur at the same time, so that we do not notice what went on. God plays a trick on us. (Well, a secular physicist would probably say it was done by nature, but the Torah tells us Who this really is!) In the language of quantum field theory, the *creation and destruction operators* of the interacting electrons make it all happen.

In a hydrogen atom, an electron appears to be continuously orbiting the proton. What is actually happening is that electrons are being continually destroyed and replaced by new ones. The same thing happens to the proton, since it is affected by the electron(s). From where do the new particles come, and to where do the old particles go? To the physical vacuum, the continuous *Tohu VaVohu* that underlies all of reality. Creation and destruction operators act upon this *Tohu VaVohu* to produce and remove particles continually, an ongoing renewal of the original creation at all times and in all places.

On their own, physicists have come to understand what the Torah tells us about the continuous renewal of creation by God. The correspondence between physics and Torah is startling. On our blackboard, next to "continuously renewed creation" we write "continual creation and destruction of particles."

The moment of creation, the big bang with its consequent *Tohu VaVohu*, the false vacuum from which all else came, the continual renewal of creation from the underlying *Tohu VaVohu*, the physical vacuum—modern physics is beginning to look more and more like the book of Genesis.

15

Everything
The Single Force Underlying All of Physics

Einstein spent the later years of his life working on two problems. The first, as we have mentioned, was to find a way around quantum theory, which he did not like. The second was to boil down all of physics into a single principle, one law that would explain everything in the universe. This law would be called the *unified field theory*. Einstein was unsuccessful in both of his quests.

Why would a smart man like Albert think that the complex world of physics could be reduced to a single equation? For one thing, he was guided by Occam's razor. As we have seen, this guiding principle is actually the scientist's first commandment: "simplify, simplify!" The goal of science has always been to obtain the simplest possible understanding of the world. Einstein was bothered by the fact that there seem to be several forces operating in nature. Several forces are certainly more complicated than one would be, so wouldn't it all be simpler if there was really only one force?

There was a precedent for this way of thinking. In the early nineteenth century, physicists wrestled with understanding two mysterious forces. The first was the force of electricity; the second was that of magnetism. Both of these are familiar to us. It is the electrical force that makes balloons stick to the wall after we rub them in our hair, and it is the magnetic force that makes compass needles point toward the North Pole. In the early eighteen hundreds, electricity and magnetism seemed to be quite different things. They behaved in different ways, and neither was understood very well.

As the century progressed, physicists began to notice some interesting things. They discovered, for example, that electrical charges could produce magnetic effects if they moved. It was possible to make a magnet by passing an electrical current through a coil of wire. This was very surprising. How could electricity produce magnetism? Weren't they different forces? It was all very puzzling.

Then, along came a man named James Clerk Maxwell. In 1871 he became a professor at Cambridge University, where he directed the Cavendish Laboratory, perhaps the most famous physics laboratory of all time.[1] Maxwell thought a lot about electricity and magnetism. He took all that was known, extended it slightly, and ended up with four simple equations (well, perhaps not all that simple). These equations[2] explained electricity and magnetism in a unified manner. That is, they showed that electricity and magnetism are not really different forces at all. Instead, there is only a single force (which we

---

[1] "Maxwell, James Clerk," in *The New Columbia Encyclopedia*, ed. William H. Harris and Judith Levey (New York: Columbia University Press, 1975), p. 1728.
[2] Which we now call "Maxwell's Equations." What else?

now called electromagnetism) that sometimes looks electrical and sometimes looks magnetic. It all depends upon how we do our experiments.

Maxwell unified electricity and magnetism. That is, he took two apparently different forces and showed that they were actually one. Maxwell had done this with two forces. Einstein wanted to do this with all the forces known to physics. How many forces were there?

Four.

We have already dealt with two of them: electromagnetism, the single force that Maxwell discovered, and gravity, which we talked about in previous chapters. We are pretty familiar with these forces from everyday live. Gravity pulls us down to earth, and electromagnetism makes our electric fans turn. Actually, electromagnetism does a lot more than that. It is responsible for the forces among atoms and molecules. It is responsible for all of chemistry. It make our bodies stick together and keeps us from falling through the floor. Every force that we directly experience in life is due to either gravity or electromagnetism.

But there are two other forces. These forces act only at distances so short we never experience them directly. Since we have no daily experience with them, we have no common names for them. Physicists call them *the strong force* and *the weak force*.

The strong force is the strongest force known to exist. It holds together the nucleus of the atom. Because of its strength, it has also been the hardest to understand. The weak force is (of course) weaker than the strong force. It is responsible for the radioactivity of many atoms, and it does a lot of other things that are harder to explain. When we compare the strengths of the four forces, we find that the strong force is the strong man and the gravitational force is the wimp. When we list them all in order of decreasing strength, we get the following:

Strong Force	(short range)
Electromagnetic Force	(long range)
Weak Force	(short range)
Gravitational Force	(long range)

Even though Einstein was not successful in uniting these four forces, physicists have continued to pursue his quest, and they are beginning to make progress. The first break occurred in the 1960s when Sheldon Glashow, Steven Weinberg, and Abdus Salam unified the electromagnetic and weak forces, showing that they are one.

Why did it take so long for physicists to accomplish this? After all, Maxwell had unified the electric and magnetic forces a century before. Part of the reason is that the weak and electromagnetic forces seem so different to begin with. They have very different strengths and very different ranges. Physicists would say that they are not *symmetric*.

What do we mean by symmetry? A Swiss psychiatrist named Hermann Rorschach gained lasting fame by taking some ink and blotting it between pieces of folded paper. He made nice symmetrical patterns. If we take a Rorschach inkblot and look at it in a mirror so that we interchange its left and right sides, it won't look any different.[3] A physicist would say the inkblot had *mirror symmetry*.

In general, when a physicist speaks of symmetry he is observing that there is some operation that can be performed on an object that will leave it looking the same as it was be-

---

[3]Apparently Rorschach did something with his inkblots other than look at them in mirrors, but physicists know nothing of this.

fore. In the case of the Rorschach inkblot, reversing left and right leaves it unchanged, so it has mirror symmetry. A child's spinning top looks pretty much the same if you rotate it about its axis, so it is said to have *rotational symmetry*.

A force can be symmetric. What do we mean by this? Suppose we consider a hydrogen atom. The electron, which has a negative charge, is attracted to the nucleus, a proton, which has a positive charge. (Recall that opposite charges attract one another.) Suppose we interchange the charges of the electron and the proton. That is, suppose the electron had a positive charge and the nucleus a negative one. The force between them would be exactly the same. We can therefore say that the electromagnetic force is "symmetric under the interchange of positive and negative charges." Physicists call this *charge conjugation symmetry*.

Physicists always assume that, on the deepest levels, the universe is symmetric in all sorts of ways. They cannot prove this, but it is a guiding principle behind all of their most fundamental theories. They are always looking for newer and deeper symmetries.[4]

Now, let's return to the weak and electromagnetic forces. They are not symmetric. If we have two particles that are attracted by an electromagnetic force and we substitute for them two particles attracted by the weak force, the force between them will not be the same. (Recall that the electromagnetic force is stronger and has longer range than the weak force).

---

[4]If you ask a physicist why this is so, he may admit that he doesn't know. If he is really honest, he will recognize that it is purely because symmetric theories are beautiful and satisfying. There is no other reason. It is refreshing to realize that cold, hard physical scientists are as driven by beauty as are artists.

If the forces are not symmetric, how could they be one and the same? How could they be unified?

There is a way. Suppose that the forces are really symmetric, but that their symmetry is somehow concealed so we do not notice it. Suppose they have a *hidden symmetry*.

To understand this, let us consider one of the symmetries of the laws of physics. If I take empty space and rotate it in some way, it still looks the same. Empty space has rotational symmetry. Suppose I put something into this space that is subject to the laws of physics. Let's say, an electron and a proton. If these particles are oriented in some particular way, say the electron is to the left of the proton, they will experience a certain force of attraction. If I now rotate them so that the electron is to the right of the proton, the force will be the same. We therefore say that the laws of physics have rotational symmetry.

Now suppose we put into the space a small, perfectly spherical drop of water. This water also has rotational symmetry. No matter how much we rotate it, it looks the same, and it behaves the same. Observing the water, it does not surprise us that the laws of physics also have rotational symmetry. But suppose it gets very cold, and the water freezes. Suppose it freezes into a snowflake.

We all know what snowflakes look like. They are very pretty, very intricate, and they always have six points. If we take a snowflake and rotate it by exactly one sixth of a complete rotation (that is, by an angle of 60 degrees), it will look unchanged. We would not be able to tell that this rotation had taken place. Physicists would say that snowflakes have *sixfold symmetry*. But suppose we do not rotate it 60 degrees. Suppose we rotate it 50 degrees, or 10 degrees, or any other angle that is not a multiple of 60 degrees. In this case, we *can* tell that a rotation has occurred, the snowflake *does not* look unchanged. When the water froze, it lost its perfect rotational symmetry and now has only a reduced sixfold symmetry. We say that the symmetry of the water was *broken*.

If all we had to look at in the universe was this snowflake, we might falsely conclude that the laws of physics themselves have only sixfold symmetry. Of course, we would be wrong. Just because the water froze and had its symmetry broken does not mean that the laws of physics changed, it's just that freezing the water *hid from us* the underlying rotational symmetry of the universe.

In the mid-1960s Glashow, Weinberg, and Salam developed a theory that showed that, deep down, the weak and electromagnetic forces are indeed symmetric. It's just that we are deluded into thinking they are not because the underlying symmetry of particles experiencing these forces has been broken. It has been broken in a manner very similar to what happened when the water froze into a snowflake.

In the case of the water, the basic rotational symmetry of the laws of physics remained unchanged. It is just that something came into existence that happened to have less symmetry. This deluded us into thinking that the universe had become less symmetric.

Glashow, Weinberg, and Salam theorized that the electromagnetic and weak forces are really symmetric, but that something appeared in the universe that was less symmetric. This caused us to be deluded into thinking that the laws of physics themselves are less symmetric. The underlying symmetry was broken, and we experience the electromagnetic and weak forces as being distinct and different. What is it that appeared and broke the symmetry? Higgs bosons. (Remember them from the last chapter?)

Now, it would be pretty complicated to explain in detail what actually happened, but we can mention the name physicists give to this event. It is called *spontaneous symmetry break-ing by the Higgs mechanism*. It is the very same mechanism, discovered by Peter Higgs, that was responsible for the ap-pearance of Higgs bosons when the false vacuum disappeared.

It turns out that the false vacuum was very symmetric. When the universe consisted only of false vacuum, the elec-tromagnetic and weak forces were not distinct; there was only a single *electroweak* force. As the universe expanded and cooled, the false vacuum froze into the physical vacuum, and the Higgs bosons appeared.

Now, when the water drop froze, the resulting snowflake could have formed in one of many different orientations. Simi-larly, when the Higgs bosons appeared and the false vacuum froze into the physical vacuum, it could have happened in many different ways. That is, there were many possible physi-cal vacuums that could have been produced. In each of them, the relative strengths and ranges of the weak and electromag-netic forces would be different. The physical vacuum just hap-pened to form in one particular way, the one in which we live. This accident appeared to break the underlying symmetry of the universe. The weak and electromagnetic forces that we observe have different strengths and different ranges, although down deep the underlying electroweak force is fully symmet-ric.

This may sound a bit like mumbo-jumbo, but in 1979 Glashow, Weinberg, and Salam were awarded the Nobel Prize for their discovery. Today, physicists regard the weak and elec-tromagnetic forces as having been successfully unified.

But physicists were not finished. They now had reduced all of physics to three forces, but this was still too many.

In 1974 Howard Georgi and Sheldon Glashow published a paper in which they tried to unify the electroweak and strong forces in a manner similar to that in which the electromagnetic and weak forces had been unified. They referred to this as a *grand unified theory* (GUT). Since this time, other physicists have suggested alternative grand unified theories (GUTs). There are many GUTs under consideration, but they all share the same main features: spontaneous breaking of an underlying symmetry by the Higgs mechanism. According to these theories, there was a time before the disappearance of the false vacuum when an even falser vacuum existed. At this time, the electromagnetic, weak, and strong forces were all one. But, as the universe cooled, this falser vacuum cooled and froze into the ordinary false vacuum, and the symmetry between the electroweak and strong forces was broken. It is hoped that future generations of particle accelerators will produce collisions of such high energy that we will be able to glimpse the underlying symmetry of these forces.

We can guess what will be coming next. Physicists are hard at work on what they call the TOE, *the theory of everything*. This theory will be the grand prize of modern physics.[5] It will unite all four forces into one. It will show that there is really only one force that exists in the universe, but that at a very early time following the big bang, the symmetry appeared to be broken so that we now think that there are a multiplicity of forces.

There is really only one force in the universe, but it is hidden. Hidden so deeply, that we think there are many. How does this relate to the viewpoint of the Torah?

The Torah tells us that there is only one true mover and shaker in the universe. The only true force is God. But, as we have seen, God keeps Himself hidden so that we can have free

---

[5]Considering the orientation of this book, I have avoided the temptation to call it the "holy grail."

will. He has done this through the process of the *tzimtzum*, which we can speak about but which we cannot readily grasp. The hidden parts of the Torah tell us that the world we perceive is but the tip of the iceberg, that there are much deeper levels of reality that are hidden from us. They speak about the layers of ever-deepening reality that underlie the world. They tell us that underlying it all is God, and only God, but that this deepest level of reality is forever beyond our grasp. Because of this, we are deluded into thinking that there are many forces operating, but there is really only One.

What we are seeing in modern physics is only the barest hint of the underlying unity of God's universe. We think there are many forces. What else could be expected of us, we who view the world through clouded glasses? But physicists are now beginning to see that there is a single force underlying it all. There is something deeper than what our senses tell us, but something has happened to hide the truth from us.

Let us return to the blackboard. We have already paired *Tohu VaVohu* with the false vacuum. We now pair the *tzimtzum* with "the spontaneous breaking of the underlying symmetry of the universe by the condensation of the Higgs boson fields." What a mouthful. This does not, by any means, sum up the *tzimtzum*, but it is at least a simple piece of hard evidence that there is much more going on than what we see before us, that there truly is but one force in the universe.

16

The Eye of the Beholder
The Quantum Theory of Measurement

In the year 1828 a chemist named Friedrich Woehler combined ammonia with cyanic acid and synthesized a chemical called urea. Urea is not a particularly complex substance. Its chemical formula is NH_2CONH_2, and it is a crystalline solid at room temperature. It has many industrial uses and is a very cost-effective fertilizer. An adult male secretes about one ounce of urea daily in his urine. Nowadays, chemistry students are not particularly excited by urea, but when Friedrich Woehler first synthesized it, it caused a revolution in all of scientific thinking.

His accomplishment was shocking because urea is an organic compound, that is, one containing carbon. These compounds form the basic chemistry of all living beings. Until the time of Woehler, all organic compounds had to be refined from biological sources, since no chemist had ever been able to synthesize one in the laboratory. Not only that, but most scientists thought that organic compounds could not be produced artificially. They believed that there was something special about life, that the processes that took place within living creatures were fundamentally different from those of the inanimate world, and that these processes could not be duplicated or even understood by mortal beings. They divided the world into two fundamentally different domains, the animate and the inanimate. Woehler's synthesis of urea radically changed this world picture.

Since this time, scientists have been continuing to whittle away at any special place that life has held in the universe. Many of life's chemical processes can now be duplicated in the laboratory. Genetic engineering is enabling biologists to experiment with the design and synthesis of entirely new organisms. Already, new strains of bacteria have been produced that manufacture insulin so that diabetics need not obtain this vital substance from animal sources. Courts in the United States have even ruled that new organisms can be patented by their inventors.

These developments have changed people's thinking about biochemical substances, the *hardware* of biology. They have shown that biological hardware is, at the molecular level, subject to the same laws as the inanimate universe. Having shown this, researchers are now beginning to chip away at the special place held by the *software* of biology, human thought itself.

Computer scientists now speak of artificial intelligence, computer programs that mimic human thought. Edward Shortliffe at Stanford University led an effort to produce a computer program called MYCIN that duplicates some of the thought processes used by doctors in diagnosing cases of bacterial infection.[1] Already, this program can diagnose diseases

---

[1] F. Hayes-Roth, D. A. Waterman, and D. B. Lenat, *Building Expert Systems* (Reading, MA: Addison Wesley, 1983), p. 20.

as well as many resident physicians. Other computer scientists have begun constructing neural networks, computer systems that duplicate the neurological structure of the brain. These systems are able to perform complicated perceptual tasks. Ben Yuhas at The Johns Hopkins University has constructed a neural network that can read lips much as is done by a deaf person.[2]

To the modern scientist, a living being has become no more than a complex machine subject to the same laws of chemistry and physics as an automobile engine. Even the brain has become nothing more than a very complicated piece of computing equipment.

Over the last 25 years, a worldwide organization of researchers has studied in detail a small worm called the *Caenorhabditis Elegans* nematode. They have constructed a complete map of its body. They know exactly how many cells are present and where they are. They have determined in detail the pattern of connections in its nervous system. They believe that, for the first time in history, they are on the verge of understanding everything there is to know about a living organism,[3] and that it turns out to be a small but very elegant machine.

Although human beings are vastly more complicated than tiny worms, many biologists believe that it is only a matter of time until they can achieve the same understanding, at least in principle, of human life and thought. Thus, the human genome project, recently begun, seeks to determine the loca-

---

[2]B. Yuhas, et al., "Neural Network Models of Sensory Integration for Improved Vowel Recognition," *Proceedings of the IEEE* 78:10 (October 1990): pp. 1658–1667.

[3]W. B. Wood, *The Nematode Caenorhabditis Elegans* (Cold Spring Harbor, NY: Cold Spring Harbor Laboratory, 1988).

tion and function of every gene in the human being. Genetic researchers hope to determine which genes cause long noses, which cause people to be prone to heart disease, and perhaps which determine musical talent. In psychiatry, researchers are looking into chemical effects on human behavior. We already know that the concentration of some very simple chemical substances in the blood—for example, sugar—can have profound effects on a person's mood. It is now believed that chemical imbalances may be responsible for major mental illnesses such as schizophrenia or depression, and it may be that even relatively common personality traits such as nervousness or compulsive neatness have a chemical origin.

It looks as if the special quality of being a human is about to be lost forever. After all, if scientists can explain all of what we are, we are no more than complex machines. The ethical implications of this are profound. If we are no more than collections of atoms and molecules, we are merely part of the world of *is*. As such, there would be no essential difference between the killing of a human being or the dismantling of a robot. Science, the ultimate cataloger of what *is* can have nothing to say about what *ought to be*.

Of course, just because science cannot tell us about right or wrong doesn't mean that there are no absolute meanings to these concepts. But this information must come from elsewhere; that's what Torah is all about. These are not new ideas to us. We have met them before in Chapter 3. But we now wish to introduce a new question. Let's forget right and wrong. Maybe (from a secular scientific viewpoint) they don't really exist. The question we wish to ask now is not a moral one, it is a logical one. Just how complete is the scientific picture of the world? Is science about to sew it all up? Are the laws of physics and biology sufficient to describe all of what we are, or is there a hole somewhere?

Many biologists would be tempted to say that there are no gaps, that ultimately scientists will be able to explain all of what exists or lives in the universe. But those who anticipate that total understanding is just around the corner are ignoring one small thing: the theory of small things, quantum mechanics.

We have already met one consequence of quantum theory. The world is an unpredictable place. We cannot say with certainty what an atom will do, and we cannot predict the actions of a thinking being. But still, from the secular scientific viewpoint, maybe that's just the way things are. Maybe things really do happen in a totally random fashion. Maybe the human mind, complex and unpredictable as it is, is not driven by free will. Maybe there really is no such thing as "mind." Perhaps this is just a word we use to describe that which we cannot predict. Maybe all human actions are no more willful or meaningful than the random decay of a radioactive atom.

It is this somewhat bleak possibility that troubled Einstein who, after all, did have at least a modest reputation for having a mind. He fought against the unpredictability of quantum theory to no avail. But as physicists grew more and more experienced with the world of small things, they began to realize that quantum theory contained within it a paradox that profoundly influenced their understanding of human thought.

It all goes back to the Heisenberg uncertainty principle and what happens when we measure anything. Let's again consider an attempt to measure the position of an electron. Suppose we perform such a measurement and determine that at a certain time the electron is at a particular point in space. What can we say about where the electron will be a short time later?

Since we have measured its position very accurately, we must be very uncertain about its velocity. Knowledge of its velocity is necessary if we are to predict where it will be in the future. Therefore, we cannot predict accurately where it will be next time we look. At the time of our initial measurement, we found the electron to be localized at a particular point; that is, we knew where it was. But at a later time, we no longer know where it is, therefore it is no longer localized. It ceases to have a well-defined position. It is almost as if the electron spreads out in space. Physicists have a way of calculating how much it spreads out. They do this using a very complicated mathematical formula called the Schroedinger equation. This equation is one of the foundations of quantum mechanics. It predicts than any localized object, be it an electron, a proton, or a bagel, will eventually spread out as time goes on. (Of course large objects, such as the bagel, spread out so slowly that we do not notice the effect during our lifetimes. This is fortunate, since otherwise our Passover cleaning would be extremely difficult.[4]) The Schroedinger equation is a very general law, and it describes the behavior of all matter in the universe. It predicts that any localized particle quickly loses its localization, and it never regains it.

Suppose we again make a measurement of the position of the electron some time later. We cannot predict in advance where we will find it, since it has spread out, but once we do make the measurement we end up finding it somewhere. Our measurement produces some answer; the electron is found to be at some location. Thus, it is again localized. But—just a minute. The fundamental equation of quantum mechanics tells us that electrons must always spread out; they can never again have a definite location. How, then, has this localization occurred?

---

[4]My wife has commented that the consumption of bagels causes me to spread out in a certain way, but discussion of this is best left to a different forum.

This paradox has a solution that is simple, but somewhat shocking. We are forced to recognize that our very act of measuring the position of the electron has localized it. Something (our observation of the electron) has happened that cannot be described by the Schroedinger equation (which says that the electron can never be localized). But—all matter in the universe is subject to the Schroedinger equation. Therefore, the act of observation involves something that is not really part of the physical universe!

This is worth repeating. All things in the physical world—electrons, protons, hamsters, and human brain cells—evolve in time according to the Schroedinger equation. This equation shows that anything that is localized in space (that is, anything whose position is known) will soon become nonlocalized, spread out in space so that it no longer has a definite position. Nothing in the physical world can stop this, since everything is subject to the same Schroedinger equation. Even if an electron is allowed to interact with a complete physical human body—eyes, optic nerve, and brain cells—the combined system of electron and human body is still described by a massively complicated Schroedinger equation. The electron must still spread out. Yet, when we observe the electron, we find it to be somewhere. Our act of observing it has caused it to be localized. Something has happened that is not describable by the Schroedinger equation, the equation which describes the entire physical world. Therefore, something exists that is not part of the physical world!

It may seem as if we are playing with words, but we are not. There is something very enigmatic going on here, and were this a more technical book we could show it mathematically. We would soon get into the extremely abstruse area of theoretical physics called "quantum measurement theory." We would run into an entirely new cast of characters with names such as Bohm, Podalsky, Rosen, and Bell. We would quickly be in over our heads in consideration of such conundrums as the Einstein-Podalsky-Rosen paradox, the hidden variable theories of Bohm and Bell's theorem. But when the smoke cleared we would be back where we started, with the conclusion that, although everything in the physical universe is subject to the Schroedinger equation, there exists something (outside the physical universe) that does not obey this equation.

This "something" is called the *classical observer*, and the existence of this observer is built into what is called the *Copenhagen interpretation* of quantum mechanics. This appellation dates from the early decades of the twentieth century when quantum theory was just getting started. Physicists in Copenhagen were the first to discover the existence of the classical observer, and the idea was so disturbing that at first the concept was considered to be just one of several alternative interpretations of quantum theory. Later, it became apparent that no other interpretations would work. In a sense, the Copenhagen interpretation of quantum theory *became* quantum theory.

What is this classical observer? It is a conscious mind. The conscious mind, through the process of simply observing the universe, causes things to happen that cannot be explained by the Schroedinger equation. We have to be careful here. The whole concept of the classical observer is widely misunderstood, and the term *classical observer* itself adds to this confusion.

In physics, the term *classical* refers to the period of physics preceding the 1920s, those innocent days before the development of quantum theory. Things were much simpler then. Particles stayed where you put them—they didn't spread out in space—and the world seemed to be deterministic and predictable. We know now that the microscopic world behaves quite differently. This is true for large objects as well, but in the world of our daily experience, the effects are so small that we never notice them. Thus, physicists continue to use the laws of the old classical physics as an approximate description of the macroscopic world.

Originally, physicists thought that the strange relocalization of a particle that occurred during measurement could somehow be explained by the fact that the measuring device, whether a piece of scientific equipment or a scientist in the laboratory, was very large compared to the thing he/she/it was measuring. It was so large that it was part of the *classical* world, while the small particle was part of the *quantum* world. Hence, they referred to the person or device doing the measurement as the classical observer.

Physicists now realize this was a mistake. There are, in modern physics, not two worlds but just one: the world explainable by quantum theory. The strange effect of the classical observer is not caused by its large physical size; it is somehow caused by its *consciousness!*

It might seem as if consciousness is not properly the domain of physics. Perhaps it is the subject of psychology, perhaps of philosophy, but certainly not physics. Physics deals with matter and energy. To the extent that the brain can be modeled as an electrical or chemical machine, physics could describe how it *behaves*, but the question of whether it has any *awareness* of itself or the world does not seem to be a question that can be answered by physics or biology.

It is true that we must use other disciplines to probe the nature of consciousness, but what modern physics tells us is that consciousness of some sort exists as a unique entity in the universe. It is not subject to the Schroedinger equation; it is not composed of matter or energy. But it does exist. And—it makes things happen. Without the presence of the classical observer, without the existence of some sort of consciousness that observes and considers what happens in the world, there would be no definite structure to the universe. It is only through repeated acts of measurement, only through repeated acts of observation, that particles turn out to be at particular locations, that chemical compounds exist, that planets and cows come to have a particular form.[5]

Physics is intrinsically incomplete. There is something else out there. This may or may not be troublesome to individual physicists, but from a Torah viewpoint it does not surprise us in the least.

It does not surprise us that humans possess consciousness, since we are able to directly experience this through introspection. Nor does it surprise us that this consciousnes cannot be described by the laws of physics, for the Torah tells us that our consciousness is not a part of the physical world. We know that there are higher levels of reality, and our consciousness derives from these deeper levels. Ultimately, our *souls* are derived from the deepest consciousness that exists, God Himself. We do have to be a bit careful here because it would be a mistake to think of God as a conscious mind such as ours. God is unknowable; we really cannot capture His existence or nature through any words we can utter or any ideas we can conceive. But just as we can think of ourselves as having conscious minds, so can we think of God in this way when we are willing to use anthropomorphic language.

Just as we realize that our modest minds are not part of the physical world, and that they can therefore cause things to happen in nature that are not describable by the

---

[5]See note at end of chapter.

Schroedinger equation, so is God's infinite "mind" not part of the physical world, and so can It cause things to happen that cannot be described by the laws of physics.

Let us recall that following the big bang, the universe existed in a state of *Tohu VaVohu*, a false vacuum in which only one force acted. We saw that there was a spontaneous breaking of the underlying symmetry of the universe. We pointed out that physicists conceive of this as having been a random quantum mechanical accident. But we have also seen that the quantum theoretical picture of the universe is necessarily incomplete. No definite state of the universe can occur, even accidentally, unless it is observed by some conscious intelligence. In the present world, there are many conscious beings who can make such an observation. But originally there was just One. The only "conscious" observer that could have interacted with the physical universe at this time was the One who created it in the first place.

Perhaps this can help us understand an enigmatic statement on our blackboard. The fourth item in our summary of the creation story is: "God used ten utterances in the creation of the world." Now, we can't fully understand what this means. We certainly know that God does not have a physical mouth and that God does not say things in the same way that we do. What then could be meant by an utterance?

A statement is not only spoken; it is also conceived in the mind. In some manner, the utterances proceeded from the mind of God. They proceeded from God's awareness of the physical universe. They were associated with His observation of His creation, and we have seen that it is only the observation of the physical universe that causes it to change and develop. The *yetzirah* of the universe, the reforming or reworking of the creation to bring it to completeness, required observation and cognizance by an intelligent consciousness, and this seems to have been a part of the ten utterances used by God to make it all happen.

There is so much here we do not understand. There is so much we can never understand. But modern physics seems to be discovering something the Torah has known all along: that without an underlying cosmic consciousness, the world as we know it could not have come into being. On the blackboard, next to the fourth item, the ten utterances, we write: "observation of the universe by a conscious intelligence."

ENDNOTE

I am being somewhat imprecise here, but in a popular text this is necessary. For those with a more technical bent, we can say the following: The act of measurement causes the wave function to jump to an eigenstate of the measurement operator. This effect contradicts the propagation of the wavefunction that would otherwise have resulted from application of the Schroedinger equation. The effect is dependent upon the act of measurement, but not upon the physical size of the measuring apparatus, since the Hilbert space of the particle's wavefunction and the Hilbert space (of very large dimension) of the measuring apparatus can be combined into a product space with a very complex wavefunction that is still subject to the Schroedinger equation. This wavefunction will never jump to a measurement eigenstate of its own accord.

With regard to molecules and chemical reactions: there will only be a certain probability that a reaction occurs. Without measurement by a classical observer, the quantum state will remain a mixture of reacted and unreacted molecules. No reaction product will definitely exist until a measurement is taken.

Finally, the so-called paradox of Schroedinger's cat is not

a paradox at all. The cat is macroscopic, but so what? It is only when I open the box and observe it that I cause it to jump to an eigenstate that is either dead or alive. From the viewpoint of the external observer, the cat is merely a physical system, and it has no consciousness. (This is a somewhat solipsistic position, but it seems to be necessary if we are to resolve the Einstein-Podalsky-Rosen paradox.)

17

Let There Be Light
Radiation in the Early Universe

One of the most enduring metaphors is the association of light with knowledge. We speak of shedding light on an issue, lament the dark ages, and are relieved that they ended with the enlightenment. Physics also finds an association between knowledge and light. It seems that most of the truly revolutionary ideas in physics have come from thinking about the nature of light.

Nineteenth-century physicists thought they had light all wrapped up. (But then again, nineteenth-century scientists thought that they had most everything wrapped up.) Light was believed to be a wave, much like the waves on the surface of the ocean. Just as ocean waves are disturbances of a physical substance, water, so were light waves assumed to be vibrations of some physical substance that pervaded the universe. This substance was called *ether* (not the kind that puts people to sleep). Ether was believed to be a mysterious substance that filled all space even though we could not directly feel or otherwise sense it. It could vibrate, and the vibrations of the ether were thought to be the light waves that we all know so well.

In this way, light was seen to be a property of a particular type of matter. This was the first of many associations between light and matter that were generally accepted. Another was that light could be produced only by matter. Without some source of disturbance, the ether would be quiet and still and all would be in darkness. It was only if matter became very hot, so that the molecules bounced around and disturbed the ether, that light could be produced. To the nineteenth-century physicist one thing was clear: without matter, there could be no light. Thus, the sequence of creation given in Genesis made no sense. How could God have first made light and only later created matter? The physicist of a hundred years ago was faced with a major *kashia* between the scientific and Torah accounts.

But time has passed, and we shall see that the *kashia* has melted away. Our understanding of light in the late twentieth century is quite different from that of the past century, and several of the cornerstones of nineteenth-century physics have vanished.

The first part to go was the concept of the ether. A famous experiment performed by American physicists Albert Michelson and Edward Morley showed that ether did not exist. Light might indeed consist of waves, but these waves and vibrations somehow occur in space itself, not in any material substance.

The second part of nineteenth-century theory to be abandoned was the concept of light as purely a wave phenomenon. As we discussed in Chapter 5, experiments have shown that light also acts as if made up of small discrete particles that we now call photons.

The final cornerstone of nineteenth-century theory to be done away with was the intrinsic dependence of light upon the presence of matter, and it is this that concerns us here. To understand this, we must first look a little bit more deeply into the nature of light.

As we have observed in previous chapters, light consists of little particles that are somehow associated with waves. What do we mean by waves and particles being "associated"? It is the waves that tell us where the particles might be found when we look for them.

We recall that the quantum theory of very small things is not deterministic. Once we know where a photon is, we cannot predict exactly where it will be found in the future. We said in the last chapter that electrons tend to "spread out" in time. The same is true of photons. More accurately, it is the waves associated with the photons that spread out in time. These waves determine all the possible places that the photon might be found when we measure its position.

This is not a property unique to light. Everything else in the universe is also made up of little particles associated with waves. For example, all ordinary matter is made up of atoms. We all learn in high school that atoms have a nucleus consisting of protons and neutrons and that electrons orbit around the nucleus much as satellites orbit the earth. When we go a little further in our study of physics, we find that those electrons orbiting the nucleus are associated with waves that determine where they are likely to be found. (These are called *standing waves* because they form fixed patterns surrounding the nucleus, unlike *traveling waves*, which move through space.) We also find that there are waves that determine the locations of the protons and neutrons within the nucleus. If we use some energy to knock an electron out of its atomic orbit, it will travel through space and be associated with a traveling wave in exactly the same way as a photon is, although it will not move as fast.

Well, if light and ordinary matter each consist of both particles and waves, what then is special about light? How does it differ from the ordinary matter we experience in everyday life? This is not a particularly important question for physicists. As long as they can understand the behavior of photons and other particles, they can develop theories of how the world behaves, and they are happy. There is no need for them to try to distinguish clearly between "light" and "matter."

But we have a different problem. We are trying to think about creation, and we know that one of the first utterances made by God was "let there be light." The purpose of our endeavor is to see whether there are any *kashias* between the Torah and scientific viewpoints of creation. To do this we must examine the scientific view of creation to see what it says about the creation of light. Was light indeed the first thing created, and did it exist even before matter came to be?

To answer this question we must first try to understand just how light differs from matter. We shall find that there is no single answer. It will depend, to some extent, on what scale we use when we look at each. We can choose the microscopic scale and study the nature of the particles that make up light and matter, or we can use a larger (macroscopic) scale and look at the overall structural features both.

Let's begin in the macroscopic world. The difference between light and matter here seems obvious. Matter comes in a number of forms. It can be solid, liquid, gas, or plasma (a form that occurs when gas becomes very hot and some of the electrons orbiting the atomic nuclei are stripped from them). All of these forms consist of either complete or partial atoms (i.e., atoms with at most a few electrons missing). Light is very different. It is a type of radiation, a collection of freely moving subatomic particles.

There are no atoms or even atomic nuclei to be found in a beam of light. But there are other forms of radiation. For example, beta rays are a form consisting of freely moving electrons. When working with the macroscopic definition, perhaps it makes most sense to distinguish between matter and radiation, and to identify light with the latter.

The other way of looking at light is to concentrate on its microscopic properties. To do this, we must look a little more closely at the nature of photons. We will see that they lack two very important properties possessed by ordinary matter: they have no *rest mass*, and they are not *conserved*.

We first consider the property called *rest mass*. Mass is the property of matter that tells us how hard it is to set it in motion and how hard it is to stop it once it is moving. Now, the theory of very fast things, special relativity, tells us that an object's mass increases as its speed increases. The lowest mass an object can ever have is the mass it has when it is not moving. This is called its rest mass.

We all know that protons, neutrons, and electrons are very small particles, and consequently they have very small rest masses, but nevertheless their rest masses are greater than zero. Photons, however, have the very strange property of not having any rest mass at all! If we were ever to stop one of them, we would find that its rest mass would be zero. Even more strangely, it turns out that this is an eventuality we do not have to worry about since the theory of special relativity tells us that we can never make them stop. Light always travels at the speed of light! (This might sound a bit obvious, but the implications of it are rather deep. Indeed, from this one physical fact alone it is possible to derive the entire theory of special relativity.) It almost seems as if photons are immaterial little bits of pure energy that go flitting about at great speed. So, unlike the protons and neutrons of ordinary matter, photons have no rest mass.

A second property of ordinary matter is that it is *conserved*, that is, although the masses of particles may be transformed into energy and vice versa, the total number of particles must remain the same.

To be more precise: particles of ordinary matter come in two types. The particles found in the nucleus of the atom are called *baryons* (from the Greek word *barys*, meaning "heavy"). In addition to protons and neutrons there are also a variety of other less-common particles classified as baryons. The electrons circling the nucleus are called *leptons* (from the Greek *leptos*, meaning "small"). There are also other types of leptons, such as *muons*, which are very similar to electrons but have a larger mass.

Baryons and leptons are conserved. This means that one baryon can be changed into another, such as when a neutron turns into a proton, but the total number of baryons must remain constant. Similarly the total amount of leptons must always remain constant. Physicists count the number of these particles by their *baryon numbers* and *lepton numbers*, meaning the total number of each. No matter what happens when the particles interact, the baryon and lepton numbers must remain unchanged.

Unlike baryons and leptons, photons are not conserved. Their number can change as they are created or destroyed. Let's consider what happens when we turn on a flashlight. The light bulb contains a filament made of metal, and when electricity passes through this filament it becomes very hot. This means that the atoms in the filament jiggle around rapidly. Atoms consist of nuclei surrounded by electrons. The motion of the atoms disturbs the nice steady patterns these electrons make around the nuclei. The disturbed electrons create photons that leave the filament to form the light that we see.

So, light appears to differ from matter microscopically in two ways: photons are not conserved and have no rest mass, while the particles of ordinary matter are conserved and do have rest mass. This seems like a clear difference. But, unfortunately, things are not quite as neat as that. There is a type of lepton, the *neutrino*, that is believed to have no rest mass. It turns out that there are other particles called *mesons* that have rest mass but are not conserved. Perhaps it is best to say that in addition to light and matter, the universe also contains some particles that lie in a gray area between them. God just did not make a universe as neatly partitioned as we might have preferred.

We now have two ways at looking at the difference between light and matter. Microscopically, light consists of photons, massless particles that can be freely created or destroyed. Macroscopically, light consists of radiation, a collection of freely moving subatomic particles. There are also gray areas, phenomena that lie somewhat between the concepts of light and matter.

Now that we understand (at least a little bit) the difference between light and matter, let's see how both came into existence. The Torah tells us that the first act of God following the initial big bang was to create light. Does this make sense in terms of current physical theories of the origins of the universe? Where does light come from? Does the modern picture of the universe still require that light come from matter?

To answer this, we must return to very first instants following the big bang. Unfortunately, there is considerable uncertainty among cosmogonists about what really happened at this time. We will therefore proceed to discuss this in a somewhat general fashion, expecting that in the next couple of decades physicists will fill in the details as soon as they learn them.

In Chapter 15 we discussed the TOE, the still sought for theory of everything that is the ultimate goal of physics. This theory, when finally developed, will show how the universe evolved from the initial false vacuum, the *Tohu VaVohu* state, which, although less than nothing, nevertheless had a very high density of energy. (Isn't physics wonderful? As a physicist, you can say things like this without people suggesting you get a psychiatric examination.) At this time the universe was perfectly symmetric—there was only one underlying force. Then, an "accidental" series of condensations occurred so that this underlying unity was obscured. An illusion occurred so that there now appear to be several forces operating, although underlying it all there is only one.

What is a force? We know that it pushes or pulls on something. Physicists would put it more precisely: a force is something that starts something moving or causes it to stop. An example of a force is the electrical force that causes balloons to stick to the wall after we have rubbed them on our hair. What is actually happening is that opposite electric charges have been induced on the balloon and on the wall, and there is an attractive force between opposite charges. Similarly, the electrical force causes the electrons in an atom to be attracted to the protons in the nucleus.

What is it about oppositely charged particles that causes them to be pulled together? Let's assume that we are talking about a single proton and a single electron. If we could observe them very closely, we would see something quite interesting. The electron would constantly be creating photons that would travel through space and be absorbed by the proton.

In addition, the proton would continuously create photons that would be absorbed by the electron. We say that the charged particles continuously *exchange photons*. It is the exchange of photons that causes the electrical attraction between them. The photons are said to *carry the electrical force*. What does the electrical force have to do with light? Everything, for even in classical physics it was known that light waves were electromagnetic in nature.

Just as the photon carries the electrical force, so are all forces in nature carried by particles that can be freely created and destroyed. The gravitational force is carried by *gravitons*,[1] the weak force by $W^\pm$ *and* Z^0 *bosons*, and the strong force by *mesons* and *gluons* (a name that comes from "glue-ons" because they make particles stick together—really!).

It is all very complicated, but originally the universe was much simpler. There was only one force. Equivalently, there was only one particle that acted as the carrier of this force. Even though the theory of everything has not yet been worked out in detail, there are theoretical reasons to believe that the original particle that started it all was a zero rest mass particle that could be easily created or destroyed. In other words, a little bit of pure energy, similar to what we now call "light." The first step in the evolution of the universe would have been the condensation of these particles out of the nothingness of the primeval *Tohu VaVohu*, and this condensation would have occurred because of a quantum mechanical fluctuation that was observed by some external consciousness. Physicists refer to the era in which this occurred as the Grand Unification.

---

[1]Gravitons are, at present, theorized; they have not yet been detected.

There are a good many details here that are uncertain. For example, what role, if any, did other subatomic particles play in the universe at this point? We have seen that, in the present universe, matter cannot be created or destroyed; baryon and leptons numbers are conserved. But this was not necessarily true in the early universe. GUTs (Grand/Unified/ Theories, which explain the underlying symmetry of the universe at this stage) allow for non-conservation of baryon number[2].

Was the product of the condensation purely photon like massless particles of light that later produced the other particles of matter? If so, this would fit nicely with our microscopic definition of light. Or, perhaps, were there also other subatomic particles at this time? This would be consistent with our macroscopic definition of light as radiation. It is a bit too early in the history of science for us to know which is correct. But we do know that, either way, the first "stuff" to fill the universe fulfills one of our definitions of light, just as the Torah tells us.

To put it another way, we can say with some degree of confidence that modern physics tells that the first instants of creation were consistent with there having been a primeval utterance: "Let there be light."

So far, we have been talking about events that happened an extremely short time after the creation, before 10^{-35} seconds had elapsed. In the next chapter, we will speak of events that happened very much later, up to a few seconds after the big bang. At this time, the universe was already getting to be pretty

---

[2]E. Tryon, "Cosmic Inflation," *Encyclopedia of Physical Science and Technology*, ed. Robert A. Myers (Orlando, FL: Academic Press, 1987), 3:733.

old. (Don't physicists have a neat way of talking?) But we will skip ahead a little in our story to discuss something that happened when the universe was positively ancient, a few hundred thousand years after the big bang.

Following the era of the Grand Unification, much of the primeval light of the universe was replaced by ordinary matter (we will see more of this in the next chapter), but not all of it. Space was filled with a mixed soup of particles and photons, all of them interacting wildly. This means that a photon could not move very far before it bumped into some other particle. Light could not travel any significant distance without being absorbed or scattered by the matter sharing space with it. The universe was opaque: it blocked the passage of light.

Following the big bang, the universe was a very hot place. At the time of the Grand Unification, there was so much energy concentrated in such a small space that the temperature was in the range of 10^{27} degrees. As the universe expanded, the amount of space increased, and the energy spread out and became much less concentrated. As a consequence, the universe cooled. By the time one second had elapsed, the temperature was already down to about 10^{10} degrees. After a few hundred thousand years, things were getting positively frigid; the temperature was only about 5,000 degrees, give or take a couple of thousand.

Before this time, atoms could not exist. The temperature had been so high that any electrons circling atomic nuclei would have been torn from their orbits. But now the temperature was low enough for atoms to form. The huge numbers of charged nuclei and electrons that had been interacting with the photons combined to form ordinary atoms, which interact very weakly with light. It now became possible for photons to travel great distances without being scattered or absorbed.

Space, which had been opaque, became transparent; light could pass through it.

In 1956 a physicist named George Gamow[3] began to think about the universe becoming transparent at this time. He realized that in physics everything has a temperature, even light. Therefore, the light that was filling all space at this time also had a temperature of about 5,000 degrees. What happened to all this light as the universe continued to expand and cool? According to Gamow, not much. Since it was interacting very little with the matter that was present, it just sort of hung around and got colder and colder as the universe expanded. Gamow reasoned that it should still be around today, but that it should be very cold with a temperature only a few degrees above absolute zero, the lowest possible temperature (-273 degrees Celsius, -460 degrees Fahrenheit), the temperature at which all atomic and molecular motions cease. This is really cold. It is so cold that light at this temperature would not be visible to the naked human eye.

Let us recall that light consists of microscopically small photons that are associated with traveling waves. Just as waves on the ocean surface have a property called *wavelength*, which is the distance between one crest and the next, so too do light waves have a wavelength. The ordinary light that we can see is made up of waves that have very short wavelengths: between 400 and 700 billionths of a meter. Such wavelengths correspond to a temperature of several thousand degrees. The colder light is, the longer its wavelength. The very cold light predicted by George Gamow would have a wavelength of about

---

[3]Gamow was also a brilliant popular writer—his books *Mr. Tompkins in Wonderland* and *Mr. Tompkins Explores that Atom* are excellent introductions to relativity and quantum theory.

a centimeter.[4] Therefore, according to Gamow, if the big bang theory is correct, all of space should now be filled with very cold light with a wavelength of about one centimeter.[5]

So, what of this invisible light? Does it or does it not exist? Before the middle of the twentieth century, there would have been no way to answer this question. The scientific equipment capable of detecting it did not exist until the 1960s.

Then, in 1965, two physicists named Arno Penzias and Robert Wilson were working at Bell Laboratories with a very sensitive antenna to develop new ways to communicate using microwave radiation. They discovered that something was interfering with their experiment. There seemed to be some other radiation that was being detected by their equipment and that was causing a whole lot of trouble. Being good scientists, they had run into situations like this before. They knew just what to do. They merely had to point their antenna in different directions until they could determine where this annoying radiation was coming from, and they could then go and turn it off. When they did this, they discovered that the troublesome radiation was coming from—everywhere! It seemed to come from the far reaches of the cosmos itself. They called it the *cosmic background radiation*. They were able to measure the temperature of this radiation, and they discov-

---

[4]Actually, things are a bit more complicated than this. There would be a distribution of wavelengths peaking at about 0.2 cm.

[5]Light of this wavelength is not usually called light in common speech. Most people refer to this very cold light as *microwave radiation*, but this is just a colloquial distinction. From the viewpoint of the physicist, there is no real difference between visible light and microwave radiation, nor do either of them differ in nature from radio waves, X-rays, or gamma rays. They are all made up of photons; they are all light.

ered that it was indeed very cold, only about three degrees Celsius above absolute zero. A group of physicists helped them analyze the situation, and then it was announced to the world: Penzias and Wilson had detected the cosmic background radiation, the very cold light that remained from the early history of the universe. They had looked backward in time and glimpsed creation.

The cosmic background radiation has now been studied in exhaustive detail. It has been probed, measured, and analyzed. It is everywhere, and comes at us from all directions equally. It fills space and comes from the farthest reaches of the universe. It is the remnant of creation that proves that there *was* a creation. It is now taken to be one of the strongest pieces of evidence for the correctness of the big bang theory.

Does the Torah say anything about this? Let us recall that there is a rather cryptic comment in the Talmud, quoted by Rashi, that tells us that God hid away some of the primal light from the initial moments of creation. It tells us that this light would be revealed again only at some future time, and that it would be used by the righteous who would be able to see from one end of the universe to the other. The cosmic background radiation is a remnant of the primal light from the initial moments of the creation. It has been hidden for all of history, and it is only now, in our time, that we have been able to see it. It does indeed extend from one end of the universe to the other.

But the cosmic background radiation is not "used by the righteous." On the other hand, it is not used by anyone—yet. We have only just discovered it. We don't quite know yet what it is good for. Will we see a use in our lifetimes? Will the righteous find within it some key to understanding what the universe is all about? We don't know the answers at this time, but the Torah has been completely accurate up to this point.

We have no reason to doubt that it will continue to be. On our blackboards, next to item 6, "some light hidden away," we write "cosmic background radiation."

18

Cooling Down
The Condensation of Matter

Immediately following the big bang, the universe was filled with very hot light. A million years later, the universe was much cooler and space was filled with a rich soup of particles and atoms. What happened in between? To answer this question, we will go into a studio and make a movie.

The hot stage lights have been turned on. The camera is in place. It is pointing at the top of a large table. On top of the table sits a large block of ice. The director calls for the camera to start. Then he calls for action. The ice begins to melt.

Let's consider this block of ice. Ice is frozen water, and water is made of molecules that consist of two hydrogen atoms and one oxygen atom. The molecules of ice are all arranged in a rigid geometrical pattern. Matter in this form is called a crystal.

Even though the molecules are arranged in a regular structure, they are not completely still. They are constantly in motion, vibrating and jiggling around slightly but never get-

217

ting very far from their appointed positions, much like a runner on first base who is trying to drive the pitcher crazy while remaining safe. Because the molecules are in motion, they possess *kinetic energy* (energy of motion). It is this kinetic energy that constitutes the type of energy we call *heat*.

At a low temperature, the molecules in the ice vibrate to a small extent. As the hot lights cause the temperature to rise, the molecules become increasingly agitated. Eventually they reach a point at which they are jerking around so violently that the organized geometric structure of the ice is broken down. No longer do they remain close to their appointed places. They now travel long distances, bumping into each other in a chaotic fashion. The crystalline structure is destroyed. The ice has now melted, and the water has become a liquid. The movie will show the block of ice turning into a puddle.

As the temperature rises further, the motion of the molecules in the liquid water becomes even more violent. Molecules begin to move so quickly that they fly off into the surrounding space to form a gas. The water has boiled and become steam.

Our movie director is not going to be able to increase the temperature of the water very much further using ordinary stage lights, but let's pretend that he has some very strong lights indeed and that he is able to increase the temperature in the room as high as he wishes. What do we see next in our movie?

The steam (water in gaseous form) is a diffuse collection of water molecules. Each of these molecules consists of two hydrogen atoms and one oxygen atom, arranged in a particular geometric structure. They are held together by chemical bonds. As in the case of the molecules in the ice, the atoms in the water molecules are not still. They vibrate slightly. As the temperature increases, their motion becomes so violent that the chemical bonds are broken, and the atoms fly apart. We say that the molecule has dissociated. At this point in the movie, we no longer have water. Instead, we have a hot gas of individual oxygen and hydrogen atoms.

As the temperature increases further, the atoms themselves break apart. Eventually, the atoms bump into each other so violently that the electrons are knocked out of their orbits, and the electrical forces are no longer strong enough to hold the atoms together. At this point in the movie, the gas of hydrogen and oxygen atoms has been replaced by a cloud of positively charged hydrogen and oxygen nuclei and negatively charged free electrons. We started with a solid. This melted into a liquid, which in turn boiled into a gas. We now have a fourth form of matter, which is called a *plasma*.

But our director is still not satisfied. He calls for more heat, and the temperature of the plasma increases further. If we could look closely at the individual oxygen nuclei, we would see that they are made up of particles called protons and neutrons held together by the *strong force*, which we encountered in Chapter 15. (The hydrogen nuclei are simpler. They are just single protons.) As the temperature rises, the nuclei in the plasma collide with such violence that the strong forces are no longer able to hold them together. At about ten billion degrees the nuclei break apart into individual protons and neutrons.

What comes next? At temperatures so high that they do not exist in the present universe, the protons and neutrons themselves break apart into their components: smaller particles called *quarks*. The temperature is now around ten trillion degrees. The very fabric of nature is becoming unglued. We recall from Chapter 15 that the electromagnetic and weak forces are actually manifestations of a single underlying *electroweak force*.

But, because of the freezing of the false vacuum, the symmetry in nature was broken and the illusion of separate electromagnetic and weak forces occurred. As the temperature in our movie reaches ten thousand trillion degrees, the false vacuum unfreezes, and the symmetry between these forces is restored. As the temperature rises still further, the symmetry between the electroweak and strong forces is restored, and eventually the temperature gets high enough for the gravitational force to enter into the process. All the forces of nature are now unified. We are in the falsest of false vacuums. We have reached a temperature that existed only a small fraction of a second after the big bang.

Wow! This has been exhausting, but now we have a movie that is worth seeing. The director has the film developed and calls us to a private showing. The film is mounted in the projector. The lights are lowered, and the movie begins. But—the director has played a trick on us. He is showing the movie backward!

At first, we observe the falsest of false vacuums. As the temperature drops, we see the freezing of this false vacuum and the subsequent freezing of the resulting false vacuums until the underlying symmetry of the universe is broken. This serves to hide the single universal force and give the illusion of four basic forces: strong, electromagnetic, weak, and gravitational. At this point in the movie, we see space filled with a chaotic soup of radiation, individual charged particles. As the temperature drops, some of these particles form into atomic nuclei, and we are left with a plasma of positively charged nuclei and negatively charged atoms. The plasma cools, the electrons combine with the nuclei to form atoms, and the atoms bind together to form molecules. We have a gas. The gas then condenses to a liquid, and the liquid freezes into a solid. We are back to our original block of ice.

What we have seen in the backward movie is similar to what happened in the universe in the era following the big bang. As we learned in Chapter 17, it all began with the primeval light. The universe cooled, and the underlying unity of the universe was hidden. The first matter then condensed out of the light: free ranging subatomic particles such as quarks. As the universe cooled further, these quarks combined into protons and neutrons (along with additional particles such as mesons).

About a minute after the big bang, the temperature had dropped to around a billion degrees and the protons and neutrons combined to form atomic nuclei.[1] There is a slight but important difference, however, between what happened in the universe and what we see in our backward movie.

In the movie, the protons and neutrons combine to form hydrogen and oxygen nuclei. In the actual universe, the protons and neutrons did not form only hydrogen and oxygen, but also the nuclei of the other chemical elements. It is possible for physicists to calculate the percentages of the various nuclei that were formed. Very little oxygen came into existence. Rather, it was mostly the light elements hydrogen, helium, and lithium that were formed. The heavier elements were produced in much smaller amounts. Physicists can now observe the universe and measure the abundances of the various elements. They match exactly those computed from their theories. This is another of the strong piece of evidence confirming the big bang.

From a universe originally filled with nothing (*Tohu VaVohu*), light came into existence ("Let there by light").

---

[1]D. Schramm and G. Steigman, "Particle Accelerators Test Cosmological Theory," *Scientific American* 258:6 (June 1988), pp. 66–72.

Following this, the underlying single force in the universe was concealed, and the illusion of a multiplicity of forces occurred (the *rekia*). This resulted in a soup of radiation and particles that cooled and condensed to form matter as we know it. We have now reached the third day in the narrative of the Torah.

In the Torah we are told that the *yabasha* ("solid stuff") appeared from among the *mayim* ("water" or "fluid"). Modern physics tells us that solid matter condensed from the soup of particles and radiation. There is a close parallel between the Torah's story and that of modern physics. Not exact, perhaps, but very close. After all, we are just beginning to think about creation.

In the next chapter we will discuss the formation of matter in greater detail.

19

Getting It All Together
Universal Gravitation and the Gravitation of the Universe

Legend has it that Isaac Newton was sitting under a tree when an apple hit him on the head. This caused him to have a sudden flash of insight into the nature of gravity. We don't know whether this story is true, but Newton did indeed have an idea that forever changed the way physicists thought about the universe. Today, we are a bit puzzled when we hear the nature of his discovery. It seems so obvious that we wonder why it took so long for anyone to think of it!

This is characteristic of many great ideas. The wheel, bifocal eyeglasses, even the safety pin—all of these advances solved problems in such a simple way that we were led to wonder why a problem ever existed in the first place. Yet ordinary minds were locked into habitual modes of thinking and were unable to break free and see the solutions. In Torah learning, this is also true. The mark of a true *gadol* (great Torah scholar) is the ability to resolve seemingly intractable *kashias* in a manner that makes us shake our heads and wonder why such a *chidush* (insight) did not occur to us.

What was Newton's insight? To understand his break-through, we must first consider the status of "natural philosophy" (the old term for physical science) in the seventeenth century.

Of course, everyone knew about gravity in those days. After all, if you dropped things, they fell down. It was also known that the earth was round and the moon orbited around it. But nobody connected these two facts. Most people assumed that the force of gravity didn't extend high enough to affect the moon. In addition, the moon was a heavenly body, and it was believed that heavenly bodies obeyed different laws of nature than those on the earth.

The situation was somewhat similar to that which existed in chemistry before Friedrich Woehler first synthesized urea. We recall (from Chapter 16) that before this accomplishment scientists generally believed that living and inanimate substances obeyed different laws of nature. Woehler was able to demonstrate that the underlying laws of nature were much simpler than had been expected—a single science of chemistry could explain both the biological and mineral worlds.

Isaac Newton did a similar thing for the world of physics. His great insight was to realize that the same force that pulls things toward the earth is also responsible for keeping the moon in its orbit. Heavenly bodies and earthly apples are subject to the same single law of nature, which he called "universal gravitation."

The law of universal gravitation is so simple that it is embarrassing to think that no one thought of it before Newton. It simply states that all objects in the universe are attracted to one other. That's it! Well, almost—he did add a bit of mathematics allowing us to calculate the size of the force, but this is just a detail. Newton's theory is called *universal* gravitation because it states that *all* objects, even apples and moons, obey the same law of gravity.

An apple falls down when we release it because the earth and the apple attract one another. The earth exerts a force on the apple that makes it fall down (that is—toward the earth). Similarly, the apple exerts a force on the earth that makes it fall up (toward the apple). Of course, since the earth is so much more massive than the apple, it is harder to make it move. Thus, the distance the earth falls upward is much smaller than the distance the apple falls downward. It is so small that we do not notice the earth moving.

The earth also attracts the moon. The force on the moon causes it to fall toward the earth. But—the moon is also moving in its orbit, as shown in Figure 19-1. As it falls (downward in the figure), it also moves sideways (to the right in the figure). Therefore, rather than hitting the earth, it misses it. It continually falls toward the earth, always missing it because of its orbital motion. This condition is called free-fall, and it is the same effect that causes artificial satellites to orbit the planet. It is the reason that astronauts appear to be weightless inside the space shuttle: both the space shuttle and the astronauts are falling around the earth together.

Now that we understand gravity perfectly (aren't we modest?), let us go back to the story of the universe's origins. We saw in the last chapter that the Torah and the scientific viewpoints are consistent: the cooling of the universe as it expanded caused the primeval radiation to condense into a soup of very dilute matter. What happens to such a soup under the influence of gravitational force?

Instead of thinking about soup, let's first consider another familiar liquid: soda pop (or, for those in New York, seltzer). It might be a good idea to perform a small scientific experiment right now. Let's go to the refrigerator and pour some

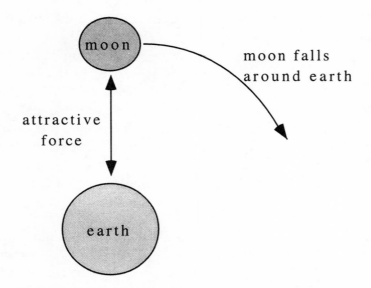

FIGURE 19–1: The moon falling around the earth

carbonated liquid into a bowl. Most any flavor will do, just as long as it has a lot of bubbles.

We immediately notice that bubbles are continually appearing at the surface of the liquid. These bubbles hang around for a little while and then pop and disappear (which explains why this drink is called what it is). Let's look a little more closely. We see that each bubble seems to deform the surface of the liquid slightly so that it rises a little bit just around the border of the bubble.[1] Now, let's watch what happens when two bubbles get close to each other. Because of the deformation of the surface, they tend to be attracted toward one another and to stick together. They then deform the surface even

---

[1]Actually, the bubbles are deforming the skin-like surface of the liquid that is formed by the surface tension.

more strongly, and attract still more bubbles until there is a little group of bubbles all united into one clump.

What is happening is that there is an attractive force between the bubbles. This force makes any random arrangement of isolated bubbles *unstable*. That is: bubbles do not like to be alone. Due to this instability, bubbles will tend to clump together in groups. If the bubbles did not pop, they would eventually clump together into large groups, and there would be very few isolated bubbles left.

Now, what does all this have to do with the origin of the universe?

Following the initial cooling of the universe, space was filled with a very diffuse collection of small particles and radiation. Due to the law of universal gravitation, all these particles attracted one another with a force that became stronger as the particles got closer together, much as do the bubbles on the surface of the liquid. Just as isolated bubbles on the liquid surface constitute an unstable arrangement, so did the collection of particles constitute an unstable arrangement due to the force of gravity. Occasionally two particles passed close by one another. The force of gravity caused them to draw near to each other. This pair of particles then teamed up to attract a third particle, a fourth, and so on. As more and more particles collected, their total mass increased so that their gravitational force became even stronger. They then attracted even quite distant particles at an ever-increasing rate. Eventually, they attracted all available particles in their vicinity. In this way, the dilute soup of matter filling the universe eventually rearranged itself into a collection of huge clouds of more concentrated matter, mostly made of hydrogen gas.[2]

---

[2]Although most cosmogonists agree with what we have just said, there is much uncertainty concerning the details of how it all

These clouds were themselves unstable once they formed, and under the attraction of their own gravity they tended to collapse inward. Additional instabilities occurred as this *gravitational collapse* proceeded, and each huge cloud broke up into millions of smaller clouds. Although these clouds were smaller than the original huge ones, they were still many times larger than our entire solar system. Astronomers refer to them as stellar *nebulae*.

Well, we can all guess what happened next. The nebulae also underwent gravitational collapse. As they collapsed, the gas of which they were composed was compressed. Now, just as gasses cool when they expand, they heat up when they are compressed. As the clouds contracted, they became so hot that they triggered nuclear reactions within them. This in turn caused them to become still hotter, until they turned into the glowing balls of gas we call "stars." In this way, the original huge clouds turned into galaxies, the immense assemblies of billions of stars that now dominate the universe.

Stars look peaceful. What could be simpler and less interesting than giant balls of glowing hydrogen and helium gas? But this is an illusion. In truth, stars are complex and turbulent structures. Within their cores, temperatures soar to millions of degrees. Violent updrafts result, and huge storms

---

happened. For example: we described the large clouds as forming purely under the influence of gravitational instabilities. Yet there had to be some initial irregularities present that started the whole process. Where did these initial irregularities come from? Some physicists suggest that they appeared very early in the history of the universe, a small fraction of a second after the big bang. Others speculate that defects in the structure of space itself led to the initial clumping together of the gas clouds. At this time, the situation is not fully understood.

occur on the stars' surfaces, many times the size of the earth. At the high temperatures of the core, nuclear reactions take place, and atoms fuse together through the same nuclear processes that cause hydrogen bombs to explode. These reactions take the simple elements of which the universe was originally made, principally hydrogen and helium, and combine them into more complex substances such as iron, silicon, and oxygen. Occasionally, stars explode, spewing their interiors into space, creating huge clouds of gas, new nebulae that merge with other clouds, undergo gravitational collapse, and then form new stars all over again.

Thus, much of the basic material of the universe has been recycled through stars many times. Initially, the universe was made up mostly of hydrogen and helium, with a few other simple elements also present. The nuclear reactions within the cores of stars converted these light elements into heavier elements. The stars acted as factories that gradually built up more complex elements and made the universe a more interesting place. Today, hydrogen and helium are still the most abundant elements in the universe, but there are also significant amounts of the heavier elements that make up our planet and our bodies.

A few billion years ago, a somewhat insignificant cloud of old, previously used gasses began to undergo gravitational collapse. The result was the smallish and rather ordinary yellow star that we call the sun. Not all of the gas condensed into the sun, however. In the process of contracting, the cloud began to rotate, and turbulent motions caused small sub-clouds to separate out and orbit the newly forming star. The light from the sun exerted pressure on these sub-clouds and blew away most of the light elements. This left smaller sub-clouds of heavier elements that condensed into the planets of our solar system, the earth included.

We now have a clear sequence in which the heavenly bodies came into existence. First, there were diffuse clouds of gas. These clouds underwent gravitational collapse, eventually forming the stellar nebulae. From these nebulae, the first stars were formed. These stars lived, and died, and in so doing gave birth to the heavy elements. Many such cycles went by, until our sun and planets were formed. And here we are today.

What does the Torah tell us about these events? Let us look at the blackboard and observe the sequence (numbers 9-12 and 14-15):

9. Fluid/vapor gathered together—a description of the cosmic gas clouds that formed first.

10. Force toward one place—clearly, gravitational collapse.

11. The first fluid bodies—the large stellar nebulae

12. The first solid matter—the planets

14. The heavenly bodies—the stars

15. The sun and the moon—our solar system.

The scientific and Torah viewpoints fit almost exactly. There is only one small problem. The sequences seem to be a little different in that the Torah speaks of the first solid matter being formed before the heavenly bodies. If, as in Chapters 17 and 18, we identify this as the appearance of the first matter from the primeval soup of particles and light, there is full consistency between the two viewpoints. But if we identify the first solid matter as one or more planets, we are left with a *kashia*.

Well, at this stage we don't have all the answers. We are only starting to think about creation. We don't understand everything yet. But when we get to Chapter 23, we will see that there is an intriguing idea that just may resolve this *kashia* for us. But—no peeking. There is something much more interesting to consider first. Notice that we did not discuss number 13 from the sequence above. To do this, we must go on to the next three chapters, those that discuss the origins of life.

20

With Fear and Trembling
The Misuse of Evolutionary Theory

A story is told about Rabbi Yaacov Kaminetsky, who was flying to Israel accompanied by his grandson. Since the plane was full, they were not able to sit together. Instead, Rabbi Kaminetsky sat next to a very pleasant and learned college professor, and they got to know each other as the plane flew eastward. Periodically, Rabbi Kaminetsky's grandson would come to see how his grandfather was doing. "Are you comfortable? Can I get you a pillow? Would you like something to drink? Can I get you your slippers?" The professor was somewhat amazed at the concern and care that the young man showed toward his white-bearded companion.

"How do you get your grandson to show you such respect?" he asked. "Mine treats me with disdain as an old-fashioned relic."

"It's very simple," said the rabbi. "My children believe that the Torah, which was given on Sinai, is the source of all wisdom in the world. As each generation appears, we get further and further from the revelation at Sinai. My children respect me because they see me as one step closer than they are to the source of truth and wisdom. On the other hand, your children believe we have evolved from the apes. They see you as one step closer to a monkey!"

Whether or not this story is true does not matter. It is still a good story. It rather brutally hits us over the head with an important point: in our secular society, people are running their lives and deciding on their moral and ethical values all based upon scientific theories. Although science can tell us nothing about what is right or wrong, about what *should* be, many modern people have no place else to look for guidance.

Let's take an example. The Torah forbids homosexual activity. It tells us that it is morally wrong. In late-twentieth-century America, there are many vocal and articulate people who tell us that what consenting adults do in private cannot in any way be immoral. One of the arguments that they give is that homosexuality is a perfectly natural situation. Look, they say, at the statistics. Over 5 percent of the population is homosexual. There may be evidence that they are born that way, that homosexuality is genetic, not learned. Even in the animal world we observe homosexual behavior among mammals. What plainer evidence could there be that there is nothing immoral about it?

Does this argument make sense? Does the "fact" that something is natural make it morally right? (We put "fact" in quotation marks since there are many reputable scientists who disagree with the observations stated above.) A little bit of thought shows us that it does not.

A number of years ago evidence appeared that there is a genetic condition, an extra Y chromosome, that leads certain men to develop a tendency towards violence. Although this evidence has since become controversial, there is speculation that there may be other genetic influences that predispose to antisocial behavior.[1] What could be more natural than a man programmed from birth to be violent? Certainly, from a scientific viewpoint, we would not criticize such a person for committing murder, would we? Why, even in the animal kingdom we see acts of violence. Researchers have observed acts of brutality and even murder among our closest relatives, the apes in Africa.

Nevertheless, we *do* condemn violence and murder. Natural, they may be. Common, they may also be. Indeed, there may even be people born with strong desires to commit mayhem. Nevertheless, no one would say that murder is morally tolerable even when committed by a person with a genetic predisposition. Certainly we must sympathize with any individual who finds himself attracted to violence due to genetic causes. Such a person would face a strong *yetzer hara* (internal urge to do something wrong), and he deserves our empathy and help. But the naturalness of his desires is irrelevant to the question of whether or not murder is morally tolerable.

Similarly, a person born with a homosexual inclination has been given a challenging *yetzer hara*. But the fact that he may have this desire does not in itself make it right for him to satisfy it. The naturalness of the desire is irrelevant to the question of what is morally correct. The Torah tells us that we are all born with our own individual *yitzrei hara* that must be mastered. *HaShem* has given us these challenges so that we may grow. *L'fum tzora agra* (according to the difficulty is the reward).[2] Indeed, let us think of opportunities for growth that would be lost were we merely to give in and go with the natural, organic flow.

---

[1]B. Kevles and D. Kevles, "Scapegoat Biology," *Discover* 18:10 (October 1997): 58–62.
[2]*Mishnah Pirkei Avos*, 5:23.

The professor on the airplane gave his children no moral message. He passed down no moral truths. All they had was a scientific theory that told them that they had evolved from inferior beings. Wasn't each generation therefore superior to the last? Wasn't this progress? Didn't this also mean that old moral ideas should be thrown out and replaced by newer more "progressive" ideas as time goes by? Isn't the human race becoming more and more advanced, ethically, as the years unfold?

Of course this is a misuse of science. But it is a common misuse. It is so common that many sensitive people in the Torah world have distanced themselves from the theory of evolution, having seen what it can do to society and to individuals. Certainly, there are many great Torah scholars who would discourage any of us from learning about Darwin's theory. It is therefore with great fear and trembling that we next tackle this difficult subject, for there will certainly be people greater than ourselves who criticize us for doing so.

Perhaps there are some who will wish to stop reading here. But, as we said in Chapter 2, this book is for those of us who are bothered by the questions that arise. It is for those of us who take Torah and science seriously and are burdened by the question of whether or not *kashias* truly exist. It is for those of us who are unable to stop thinking about creation.

And so we go on. But we keep in mind this very important truth: evolutionary theory, whether true or false, is only a scientific theory. It can only tell us something about how the world *is*, not about how it *should be*. Even if it is found to be true that man evolved from simpler, more *physically* primitive creatures, it is also equally true that an event happened at Sinai to a group of people who were on a far higher *spiritual* level than we, and that all that we know about right and wrong was transmitted to us by the people of that generation. In the words of Rav Avraham Yitzchok HaCohen Kook:

> It makes no difference for us if in truth there was in the world a Garden of Eden during which man delighted in an abundance of physical and spiritual good, or if the actual existence began from the bottom upwards, from the lowest level of being towards its highest, an upward movement. We only have to know there is a real possibility that even if a man has risen to a high level, and has been deserving of all honors and pleasures, if he corrupts his ways, he can lose all he has, and bring harm to himself and to his descendants for many generations, and that this is the lesson we learn from Adam's existence in the Garden of Eden, his sin and his expulsion.[3]

Now certainly Rav Kook is not saying that the story of the Garden of Eden is not true. But he is saying that truth or falsity of the story, the question of what *was*, is as irrelevant to the purpose of Torah as is the truth or falsity of evolutionary theory. The Torah has no intention of teaching us science. It deals only with what *ought to be*.

So let us bear in mind, as we consider evolution, that it is only a scientific theory. Let us remember that it cannot teach us anything about right or wrong, no matter how it is misused by others. Let us keep it in its place and be as dispassionate as possible toward it. Finally, let us begin . . .

---

[3]T. Feldman, *Rav A. Y. Kook, Selected Letters* (Ma'aleh Adumim, Israel: Ma'aliot Publications, 1986), p. 12.

21

Let There Be Life
The Origin of Life

The origin of life is the central enigma of creation to both the rabbi and the atheist. But before we can deal with the question of how life came to be, we must first understand what life is. Why is it that one cell is alive, while another is dead? How is a virus different from a molecule of sugar? Why are mushrooms different than the dead wood on which they grow?

At one time, natural philosophers believed that the difference between living things and inanimate objects was easy to understand. Living things could move by themselves, inanimate objects couldn't. This was an earlier and more innocent era in which machines such as the automobile had not yet been invented. No one had ever seen a device that could move by itself. Today, this all seems rather quaint.

In more recent times, modern natural philosophers (who are now called scientists, even though many of them still earn a doctor of philosophy degree) have suggested that the essence of life is the ability to reproduce. Thus, groundhogs produce other groundhogs, and viruses produce copies of themselves (although they have to infect other life forms—cells—to do it). Thus living organisms are essentially self-reproducing systems.[1] This seemed to be a pretty good definition of life until the 1940s when a mathematician named John Von Neumann began to wonder whether humans could design and build artificial machines that would reproduce themselves. He discovered that the answer is yes, and he then went ahead and showed how to do it. Along the way, he pretty much invented the modern digital computer. Nowadays, self-reproduction is no longer a mystery. Any serious user of computers will tell us that he fears computer viruses, programs that reproduce themselves and spread from machine to machine. Self-reproducing systems are now well known, yet we do not consider them to be alive.

At the moment, biologists don't really seem to have a good definition of life. It is a pretty gray area, and there is a good chance that this lack of definition will cause all sorts of ethical and legal conundrums in the years ahead. (Suppose, for example, someone builds a robot that can reproduce itself and produce offspring. If a person turns one of these off, is he guilty of murder? We can imagine the Supreme Court case: X75B vs. Ginsberg.) But this issue need not concern us here. Even though we may not know exactly what life *is*, biochemistry has given us the ability to discuss how life *works*.

All living things are made largely of proteins. These are very large and immensely complicated molecules often containing hundreds of thousands of atoms. Yet, as complex as they are, as varied as they are, proteins are amazingly

---

[1]It has even been suggested that a cat is nothing more than a machine that makes new felines from tuna fish and milk.

simple! How could this be? Let us consider the works of Shakespeare.

The various plays and sonnets are certainly complex, yet they can all be decomposed into sequences of just 26 different letters. This very simple alphabet of only 26 letters gives forth the sublime complexity of the bard's writings. Proteins, too, can be decomposed into sequences of simple "letters." All proteins are composed of 20 different elementary compounds called *alpha-amino acids*.[2] To make a protein all we have to do is take some of the 20 amino acids and string them together in a well-defined sequence that is called a *polypeptide chain*. We are taking the 20 letters that comprise the language of life and writing out a long word that specifies the protein we want. The polypeptide chain then does the rest all by itself: chemical forces cause the linear structure to fold and bend until it takes on the unique convoluted form of that particular protein.

This is what happens in our bodies and in the cells of all living things. The same amino acids are used, but different protein "words" are spelled out. How do the cells know which proteins to make? They have a recipe book, a list of words that are stored in the nucleus of the cell. They just read the words and make the proteins to match them. The words on this list are called "genes," and the list that contains them is called *DNA*, one of the most beautiful and fascinating substances known to man.

DNA stands for deoxyribonucleic acid. It is called a nucleic acid because it is found in the nucleus of the cell. For some time it was known that DNA lay within the nucleus of

---

[2] "Genetic Code," *McGraw Hill Encyclopedia of Science, 7th Edition*, ed. Sybil Parker et. al. (New York: McGraw Hill, 1992), 7:675–678.

the cell, and that it somehow had to do with the encoding of genetic information. Just how it all worked, however, was not known until 1953 when John Watson and Francis Crick, working at Cambridge University, discovered the structure of this remarkable substance.[3]

Just like proteins, DNA is both complex and simple. Crick and Watson discovered that it is composed of a double helix: two long linear strands that coil around each other like the strands in a rope. Each of these strands is composed of a sequential arrangement of simple substances called *bases*.[4] Altogether, there are four different bases involved: *adenine, cytosine, guanine*, and *thymine*. These are usually denoted by the letters A, C, G, and T. Thus each strand of DNA is a long word made up of these four letters, for example ACCGTATGCA. . . .

Let us recall that the DNA strand contains the list of words that are the recipes for the proteins to be made by the cell. The way this works is that each group of three bases is actually a code representing a particular amino acid. For example, the sequence ACC indicates the amino acid threonine. Whenever the sequence ACC occurs in the DNA, it indicates that a molecule of threonine should be added to the protein that is being produced. The association between the amino acids and the triplets of bases is called the *genetic code*. This

---

[3]We highly recommend that the reader look at Watson's book *The Double Helix*, which chronicles the discovery of the structure of DNA, not only for the scientific information but even more so for the very human story of just how scientists really function in their lives and work.

[4]Actually it is a little more complicated than this. The bases are held together by sugars and phosphate groups, but these are relatively uninteresting.

code is universal. That is, the same code applies to all animal and vegetable life. When the sequence ACC occurs in a squid, it represents the same amino acid as in a turkey or a turnip.[5]

How does the organism reproduce? It is here that things really get clever. As we mentioned, DNA is a double helix. What holds the two strands together is the attraction that the bases have for each other. But guanine is attracted only to cytosine, and adenine only to thymine. Thus, a C in one strand sticks to a G in the other, and an A to a T (see the upper part of Figure 21–1). Each of the two strands of DNA is very similar to the other, but it is sort of a complement, with A substituted for T, C for G, and so on.

When the cell divides, the DNA splits apart into two separate strands. Both strands contain the same information, and the cell then uses each as a pattern to make a new complete double helix of DNA. This process is shown in Figure 21–1. One DNA molecule (shown at the top) has given birth to two perfect copies of itself (shown forming at the bottom)—we have one of John Von Neumann's self-reproducing machines.

Now, how did all this come to be? Where did the first DNA come from? At one time, biologists thought they knew the answer; now it is no longer so clear.

The answer seemed to come in 1953 when a researcher named S. L. Miller took a glass globe filled with water, meth-

---

[5]The astute reader will have noticed that there are 64 possible sequences of three bases, yet only 20 amino acids. This is because certain distinct sequences of bases are synonyms, that is, they represent the same amino acid. Thus, threonine can be represented by ACC, ACU, ACA, or ACG—"Genetic Code," *McGraw Hill Encyclopedia of Science, 7th Edition*, ed. Sybil Parker et. al. (New York: McGraw Hill, 1992), 7:676.

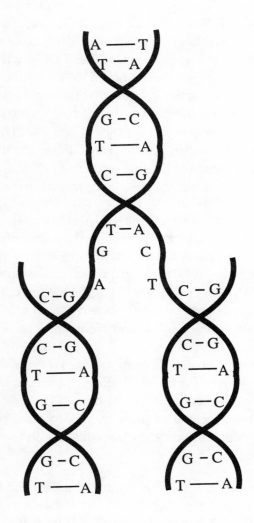

Figure 21-1: The replication of DNA

ane, ammonia, and hydrogen and made electrical sparks within it.[6] The substances present in the globe were those believed to have been present a couple of billion years ago when the earth's oceans were still young. The electrical sparks duplicated the effects of lightning. Miller analyzed the contents of the globe after the experiment was finished and found that some simple amino acids had been formed by the "lightning."

Aha! That was how it happened! The earth's oceans contained some primitive substances that, under the influence of lightning, formed amino acids. These amino acids accidentally (!) bound together into varied polypeptide chains, some of which formed crude proteins. Other simple substances, bases, sugars, and phosphates, accidentally formed long linear chains that became DNA molecules. Some of the proteins were able to cause the DNA molecules to split and reproduce themselves, and some of the DNA chains were able to make the proteins that caused this to happen. The first simple life had appeared.

In 1953 this was a beautiful picture, but today it is no longer satisfying. For one thing, the experiments of Miller do not tell the whole story. It is true that electrical discharges can make amino acids, but the acids so produced are not stable in the presence of oxygen. At present, very little is actually known about the composition of the oceans at the time all this was supposed to have happened,[7] but there seems to have been too much oxygen around for life to have appeared in this way. Some evolutionary biologists now speculate that life might have begun deep in the ocean, near underwater volcanoes that rid the waters of oxygen. Others suggest that life did not begin in the oceans at all, but rather within clay (*"and*

---

[6] "Origin of Life," *McGraw Hill Encyclopedia of Science, 7th Edition*, ed. Sybil Parker et. al. (New York: McGraw Hill, 1992), 10:45.
[7] Ibid.

the HaShem Elokim formed man from the dust of the earth," Genesis 2:7).

But there is another, more fundamental problem with this explanation of the origin of life. This has to do with the accidental nature of how it supposedly happened. Not only did amino acids have to form, but they had to accidentally bind together in the exact sequences necessary for useful proteins. A protein only 50 amino acids long (about as complex as insulin, a fairly simple protein) can be assembled in 10^{65} different ways, and only a minuscule fraction of the possible sequences have any utility at all. Most proteins are much larger than this. Accidental production of a useful protein is about as probable as giving a monkey a typewriter and having him peck out *Hamlet*. Not only would proteins have to come into being accidentally, but DNA would have to be produced by a still-undiscovered mechanism to encode their structures, and the proteins necessary for splitting and reproducing the DNA would have to be produced along with it. Certainly, accidents can happen and, given enough time, they might. But, how long is enough time? Recently mathematicians have begun to calculate just how probable it is that life could have originated in this way in the billion or so years available. Now, a billion years is a long time, but the probabilities of the accidents happening are so small that there just wasn't nearly enough time for it all to have happened this way.[8]

To get around this problem, some researchers have taken the dramatic step of suggesting that life *did not* originate on

---

[8]L. Spetner, "Natural Selection: An Information-transmission Mechanism for Evolution," *Journal of Theoretical Biology* 7 (1964): pp. 412–429. Also, L. Spetner, "Information Transmission in Evolution," *IEEE Transactions on Information Theory* IT–14:1 (January 1968): pp. 3–6.

earth but rather that it developed elsewhere in the universe (by a mechanism they do not explain) and was brought to the earth by meteorites or comets. This idea was originated in 1908 by the Swedish chemist Svante Arrhenius, who suggested that living spores from outer space seeded the earth.[9] For many decades this theory was not taken seriously, but in the 1980s Francis Crick, who along with Thomas Watson had won the Nobel Prize for discovering the structure of DNA, formalized the theory into one now called *cosmic panspermia*, and the prominent physicist Fred Hoyle lent his support to the idea. This theory may still seem a bit far out (literally) but evidence such as the discovery of quasi-genetic material in meteorites[10] has given it some respect. In addition, discovery of possible life forms in a meteorite that may have come from Mars[11] raises the possibility of seeding of the planets by more complex genetic materials.

Of course, cosmic panspermia does not answer the question of just how life might have originated in outer space. Therefore, it merely substitutes one problem for another. Curiously, many scientists don't seem to be bothered by this.

Did life originate in outer space? If so, science is remarkably consistent with the Torah that tells us that genetic material was formed (on the third day) before the heavenly bodies (on the fourth). Did the genetic material rain down on the earth and enter clay to form the first terrestrial life (consistent with man's origin from the dust of the earth)? Did this genetic material lie dormant until watered by the rains that

---

[9] "Life," *Colliers Encyclopedia*, ed. W. Halsey and B. Johnson (New York: Macmillan Educational Company, 1989), 14:622.
[10] Web site http://seti1.setileague.org/articles/simple.htm, Dec. 1997.
[11] *Aviation Week and Space Technology*, 145:7 (August 12, 1996): 24–25.

developed as the earth cooled, and did it then spring forth as the early vegetation that proceeded animal life? This would be highly consistent with what the Torah seems to tell us (numbers 13, 22, 23, and 24 on the blackboard). The problem is that scientists have not yet agreed on the sequence. So we do not yet have consistency between the Torah and scientific viewpoints on this issue. On the other hand, we do not have any *kashias*. We will just have to wait a bit longer until it is all worked out.

Meanwhile, on our blackboards next to "13. genetic material," we write: "panspermia"; and next to "23. man from the dust of the earth," we write "life from clay." We write both of these with question marks since the scientists have not yet settled the issue. We will just have to wait as they continue to think about creation.

22

A Monkey Is His Uncle
The Theory of Evolution

In one way or another, life appeared on earth. But it was extremely simple life—just the bare bones—DNA, some proteins, and a mechanism for the DNA to reproduce itself. Since then, things have gotten a bit more complicated. In our era, we have carrots, wombats, and your Uncle Sidney. How did we get from one state of affairs to the other?

An answer was suggested by a man named Charles Darwin, and the world has not been the same since. Charles Darwin was born in 1809 and studied medicine at the University of Edinburgh and theology at Cambridge.[1] Between the years of 1831 and 1836 he traveled as official naturalist aboard a ship called the Beagle and studied the various life forms on the islands they visited. He was struck by the beau-

---

[1] "Darwin, Charles Robert," *The New Columbia Encyclopedia*, ed. W. H. Harris and J. S. Levey (New York: Columbia University Press, 1975), pp. 721–722.

tiful way in which animals seemed to be adapted to the particular environments they lived in, and this inspired him to formulate his *theory of evolution* to explain how this came about. The theory was independently discovered by Alfred Russel Wallace, and the two men published their ideas simultaneously in 1858.

Like all good theories, Darwin's is actually very simple, and it seems to explain a lot. Just how good a theory it is, just how much it is able to explain, we will discuss later. For the moment, let us content ourselves with what the theory is. Darwinian evolution can be boiled down to two basic ideas:

1. Living things gradually change and evolve into new species.

2. This is caused by two effects: random mutation and natural selection.

We will start by discussing mutation. In Chapter 21 we described the replication (reproduction) of DNA. We showed how one molecule of DNA gives birth to two exact copies of itself. Actually, this does not always happen. Nothing in the physical world is perfect, and occasionally errors occur while copying the spelling of the genetic words in the original DNA. As a result, altered versions of the original DNA are sometimes produced. Another way in which altered DNA can occur is through disturbances of the DNA molecule due to radiation or certain chemicals. When a new variant of the original DNA appears, we say that the genetic material has *mutated*, and the change is called a mutation.

It is a fundamental tenet of Darwin's theory that mutations are completely random. What is the effect of such a mutation? The altered DNA no longer spells out the same proteins as the original. One or more of the letters in the word lists has changed, and new protein-words result. In most cases, these new words will be pure nonsense.

As an analogy, suppose we take an English word such as CARROT. If we change one of its letters, quite at random, we will probably get a word such as CARBOT, CARROM or CXRROT. All of these are nonsense. It is only if the letter C changes to a P that we will get the meaningful word PARROT. There is only about one chance in 150 that this will happen (assuming all changes of all letters to be equally likely). If the changes are more extensive than one letter, it is much less likely that we will get a meaningful word (as low as one chance in 40,000 if all letters are changed).[2]

The words within DNA that spell out the proteins are very much longer than our six-letter English word. Many of them are hundreds of characters long. It is very unlikely that a mutated word in DNA will spell out a meaningful protein. It is even less likely that any proteins so produced will be in any way useful to the organism. It is most probable that the mutated DNA will produce a life form with proteins so messed up that it does not survive. Occasionally, however, *very occasionally*, a new protein will appear that alters the organism in a beneficial way.

Let us take an example. Consider a population of Africans with dark skins. Of course, people seem much more complicated than the minimal life forms we have been talking about, but from a genetic viewpoint this difference is superficial. Human bodily structures are determined by their DNA and proteins just as are those of the simplest life forms. It is really a matter of quantity, not quality.

---

[2]There are 308,915,776 possible 6-letter combinations. An informal Monte Carlo study of *Webster's Ninth New Collegiate Dictionary* produced an estimate 7,700±3,900 words with 6 letters.

Africans have dark skins since the DNA in their bodies programs their cells to manufacture proteins that in turn cause their skins to produce large quantities of melanin, a dark brown pigment. Occasionally, a dark-skinned person will give birth to a child whose DNA has mutated so that his skin is not as dark as his parent. Central Africa lies close to the equator, and there is a lot of sunlight throughout most of the year. Since the climate is tropical, people spend many hours outdoors. The melanin in Africans' dark skins protects them from sunlight. The light-skinned child, however, does not have as much protection, and there is an increased chance that he will develop a skin cancer from the effects of the sun. If he does, and he dies before he is able to have children, his mutated DNA will not be passed on to a new generation.

Suppose, however, that some Africans migrate to Sweden (as anthropologists speculate actually happened over a period of time). Sweden is a Northern country without much sunlight. It is also cold, and people wear heavy clothing and spend a lot of time indoors. In this environment, our light-skinned person will probably survive to pass his DNA on to his children. On the other hand, the dark-skinned people who came with him will be at a disadvantage. To lead a healthy life, we must have a supply of vitamin D. Before the advent of vitamin D-enriched milk, most people got their supply from their own skins, which produce this vitamin when exposed to sunlight. In the northern climate, the light-skinned person would produce sufficient vitamin D, but the dark-skinned people's skin would be shaded from the sun and they would suffer from vitamin D deficiencies. Many of them would die before they could have children, and the DNA for dark skins would disappear from the population. After a few hundred thousand years (no one is sure how long it would really take), the entire population of Sweden would be light skinned, very well adapted to its environment.

This adaptation, which is caused by the increased survival of individuals genetically suited to their environment, is called *natural selection*. Sometimes we call it "survival of the fittest" (another scientific principle commonly misused, often by lawyers, financiers, and presidential candidates).

Natural selection is a very simple yet ingenious mechanism for change. There is little doubt that organisms do indeed experience mutation and natural selection. In our lifetimes, biologists have seen bacteria that could be treated by antibiotics such as penicillin evolve so that they have become resistant to these drugs. Drug-resistant tuberculosis microbes are now a serious health problem in many cities.

Another case in which natural selection has been observed is in the evolution of Norway rats. For many years, exterminators have used a poison called warfarin to kill these rats, and it has worked pretty well. But along the way a few rats appeared that, due to mutations, were immune to this poison. They survived and their offspring inherited this immunity, while the nonresistant rats were killed off. The species has now evolved to such an extent that exterminators must use a different poison when trying to exterminate rats in many locations.

So, evolution by natural selection does occur. But Darwin said more than this. He said that natural selection is the *only* mechanism by which one species evolves from another and that *all* the changes that have taken place since the first life appeared were caused by this mechanism alone. Darwin was a good scientist, and as a scientist he took his theory and made predictions that could be tested. He predicted that when paleontologists, those practitioners of the then-new study of fossils, began to dig more deeply into the earth, they would find petrified evidence of the gradual change of one species into another. They would see that all life evolved slowly, and that new species came into being purely through mutation and natural selection.

It is now over one hundred years later. What is it that researchers have found as they looked into the fossil record? Did they find a slow drift of one species into another as Darwin predicted?

They did not.

What paleontologists did find was quite unexpected. They observed that species were remarkably stable over very long periods of time, that species did not appear to evolve slowly, but rather that old species seemed to disappear while others came into existence quite suddenly. Although this was a shock to the world of science, it was not to the world of Torah. Indeed, in 1908 Rav Kook predicted:

> In general the idea of gradual evolution is also only in its beginning, and there is no doubt that it will change its form and give birth to conceptions that will also include sudden leaps to complete the picture of nature, and then the light of Israel will be understood in its very clarity.[3]

Does this mean that the theory of evolution is dead? Not at all. It is quite common for scientific theories to need adjustment. Seldom does a theory spring forth in its final form. There are paleontologists hard at work trying to understand how random mutation and natural selection could produce the sudden jumps that are seen in the fossil record.

---

[3]T. Feldman, *Rav A. Y. Kook, Selected Letters* (Ma'aleh Adumim, Israel: Ma'aliot Publications, 1986), pp. 13–14.

In 1972, the prominent researchers Steven Jay Gould and Niles Eldridge proposed a modification of Darwin's theory called *punctuated equilibrium*, which attempts to explain the sudden appearance of new species.[4] They suggested that drastic changes in the environment over short time periods, perhaps in localized areas, are the triggers of change. These environmental changes when coupled with mutation and natural selection could lead organisms to readapt to a modified world quite rapidly.

There is, however, a problem with their theory. A theorist named Lee Spetner has developed a mathematical theory that determines just how fast all of this can happen.[5] He found that the rate at which evolutionary changes can occur is very low. That is, Darwin was correct in stating that his theory predicts a slow gradual metamorphosis of one species into another. Evolution must happen slowly, yet it is observed to happen quickly. There is an internal *kashia* within the theory of evolution!

At this time, there is no consensus in the scientific community as to whether or not punctuated equilibrium is a good theory. There is no universally accepted theory of why the fossil record does not conform to Darwin's predictions. Indeed, the principal evidence given for the accuracy of evolutionary theory is not the fossil record, but rather the astound-

---

[4] "Organic Evolution," *Colliers Encyclopedia*, ed. W. Halsey and B. Johnson (New York: Macmillan Educational Company, 1989), 9:481.

[5] L. Spetner, "Natural Selection: An Information-transmission Mechanism for Evolution," *Journal of Theoretical Biology* 7 (1964): pp. 412–429. Also, L. Spetner, "Information Transmission in Evolution," *IEEE Transactions on Information Theory* IT–14:1 (January 1968): pp. 3–6.

ing similarities in structure between different species. Thus, a giraffe has basically the same neck structure as a mouse, and the insulin proteins found in many species are remarkably similar to one another. It is assumed that this similarity is present because they all evolved from similar ancestors.

We, of course, could advance a different explanation. We could point out that the similarity in structure could be due to their having the same creator, much as a Chevrolet and a Pontiac look alike under the hood since they have both been designed by General Motors. We could also point out other defects in the reasoning. For example, we could show that there are statistical problems with evolutionary theory as it is presently formulated. A major difficulty is the probability that the entire evolutionary sequence could have occurred in the time available. It is a fundamental tenet of Darwin's theory that mutations occur accidentally. There is no meaning to the whole process; there is no guiding hand. Under such circumstances, the probabilities of the required changes taking place, even over billions of years, are quite small.[6]

But just as Torah scholars are able to tackle *kashias* without concluding that Torah is invalid, we cannot assume that a few difficulties will be fatal to evolutionary theory. We will

---

[6]This is an oversimplified statement of a complex subject. For example, the work of L. Spetner shows a statistical inconsistency between the mechanism of random mutation/natural selection and the concepts of parallel and convergent evolution [L. Spetner, "Natural Selection versus Gene Uniqueness," *Nature* 226 (June 6, 1970): pp. 949–949]. This is what we mean by saying that there is too small a probability of a "required change" taking place. (The example that Spetner gives is the vanishingly small probability that the vertebrate and invertebrate eyes could have come out so similar when they supposedly evolved independently of one another.)

have to wait a little bit longer to see whether and how these problems are resolved. It is not our purpose here to discuss the validity of a scientific theory. It is rather to ask the following question: Suppose it turns out that the problems in evolutionary theory are resolved. Suppose in a few more years Darwin's theory, or some slightly modified version of it, is confirmed to such an extent that all reputable scientists accept it as true. Would this constitute a problem for the literal truth of the Torah?

It would not. Let us recall, from our blackboard, that biological man was created in a manner *consistent with the laws of nature*. The Torah does not tell us what these laws are, only that a biologist observing the events would not say that anything happened out of the ordinary. If evolutionary theory is correct, then this *is* the law of nature, and, from a Torah viewpoint, this is the way in which the biological being we call *Homo sapiens* would have come to be on this planet.

There are, however, two points in which the Torah story differs from the tale told by biologists. Neither of these creates a *kashia*, neither of them represents an inconsistency between the two stories. It is just that the Torah emphasizes two points that do not lie within the realm of science.

The first point has to do with the randomness and meaninglessness of the evolutionary process as viewed by the scientist. The Torah tells us that nothing in this world is meaningless, that nothing is random. Meaning may be hidden from us, and events may be beyond our ability to predict—so they appear random—but the hand of God is always acting to drive the world toward the purpose for which it was created. The purpose of creation was man, and God created a universe whose physical laws assured that all the steps of evolution would move inexorably toward this goal. This is not subject to scientific verification, but neither does it contradict science. Science is silent on the subject of meaning.

The second point that the Torah makes is that the appearance of the first true animate life was not entirely explainable by the theory of evolution. Of course, the evolution of the animals' physical bodies could be so explained, but the Torah tells us that animals are more than just self-reproducing machines—they have animal souls. The appearance of the first animal souls was a new creation by God (Genesis 1:21, item number 17 on our blackboard). Again, this does not in any way contradict the scientific picture. The soul, be it human or animal, is not observable by the methods of science.

So man came into existence. How long ago did this happen? At present, the answer is unclear. There are many fossilized primitive beings, such as Neanderthal and Cro-Magnon man,[7] vying for the title of ancestor to us all, but no transitional fossils that show which one turned into *Homo sapiens*.[8] It does seem safe, however, to pick a date between 300,000 and 30,000 years ago as the birthday of the biological ancestor of Charles Darwin.

As we have seen in Chapter 9, one interpretation of the Torah tells us that *HaShem* created the first man by placing a human soul in the body of an existing living creature, and that

---

[7]At the present time, Cro-Magnon man is the favored candidate. It is currently believed that Neanderthal man came to a dead end. This leads to an interesting speculation. Cro-Magnon man was smaller than the co-existing Neanderthals. Could the brutish Neanderthals have been the giants that are present in the oral traditions of many cultures?

[8]"Evolution of Man," *Colliers Encyclopedia*, ed. W. Halsey and B. Johnson (New York: Macmillan Educational Company, 1989), 9:487–488.

there were 974 generations of biologically complete human beings before this. If Homo sapiens first came into being 30,000 years ago, and if *Adam*, the first true man, were created nearly six millennia ago, then there were about 24,000 years of biological Homo sapiens without human souls. This works out to 974 generations if a generation takes about 25 years, quite a reasonable number.

What if Homo sapiens appeared earlier, say 300,000 years before *Adam*? If we divide this time span by 974 generations, we find that each generation lasted around 300 years. Long life spans are reported to have occurred early in human history. We see this throughout the first chapters of Genesis. It was not at all unusual then for people to live hundreds of years. But today, people rarely live to be one hundred. Are long life spans consistent with the laws of human biology as we know them?

Why do people die? Some, of course, die from accident or disease, but many seem to just wear out as they get older. At first glance, this does not surprise us since it is also true of just about all the other things we know of: cars, refrigerators, wind-up toys, and trousers. But there is a significant difference between a biological organism and a pair of pants. If we get a hole in our slacks, it remains a hole unless someone comes along with a needle and thread. If we cut our finger, it heals by itself. The organism is alive, and its cells are constantly reproducing and repairing damage as it occurs. So why does it wear out?

Biologists have found that cells act as if they are programmed to divide and reproduce a certain number of times and then stop. There seems to be information within the nucleus that tells the cell what this magic number is. When we are young, our cells divide quite readily and bodily damage is easily repaired. If a young child looses a finger tip, the body is able to regenerate a new one, fingernail and all.[9] After roughly the age of twelve, the cells do not reproduce as well and the body loses this ability. We become elderly because many of our cells have stopped reproducing and repairing damage. Wounds take a long time to heal. Regenerative mechanisms fail to operate. Damage accumulates, and we wear out and die.

What if we could change the information in the cell that controls the number of times it can divide? What if we could double the number? Our bodies would retain their youthful vigor for twice the number of years, and we would tend to live much longer. This is not just a theoretical possibility. In the laboratory, scientists can change the programs of certain cells to allow them to reproduce indefinitely. They even call this procedure "immortalization." Indeed, cancer cells seem to have the property of immortality. Therefore, it is entirely consistent with biological theory to imagine a person who could live a few hundred years.

We might speculate that in the early days of our species, people were programmed to live lengthy lives. Up through the time of the patriarchs this seemed to be true. But then something happened. A mutation occurred that limited life spans, and nowadays we must settle for 120 years at the most. There are many unanswered questions. Why would such a mutation have become established through natural selection? Did shorter-lived individuals reproduce more quickly, and eventually overwhelm their more leisurely neighbors with offspring? We cannot be sure, but we can state with some certainty that there is no scientific problem with assuming that

---

[9]K. J. Rose, *The Body in Time* (New York: Wiley Science Editions, 1988), p. 66.

the early generations of Homo sapiens took as long as 300 years.

Perhaps we can now understand a very enigmatic portion of the Torah, the attempt by *Adam* to find a mate among the other animals. How could we possibly think that such an exalted person could find true happiness with a duck? Suppose, however, that *Adam* did not live totally alone. Suppose that there were plenty of people around but that they did not have souls, neighbors that were biologically complete but spiritually incomplete. It is quite understandable that the first true man would search for a mate among these creatures, but find that he could not encounter a partner that would share his quest for spiritual growth.[10]

The Torah reports that *Adam* was an hermaphrodite, a being that was both male and female. This does not at all contradict the laws of biology. Even today, we occasionally find that people are born with both male and female organs, although it is quite rare and we regard it as a birth defect. This phenomenon is recognized in *Halacha* (Jewish law), in which such a person is called an *androgynus*. There are no problems here. But the Torah then goes on to tell us something that is certainly *not* encountered nowadays. After this person's unsuccessful attempts at dating, *HaShem* caused him/her to fall into a deep sleep, and he/she was then split apart into separate male and female beings. This is pretty major surgery, even with deep anesthesia. It is certainly not in accord with the laws of biology as we know them.

But this does not create a contradiction. As we have seen above, the nature of a *nais she lo b'derech hatevah* (miracle outside the laws of nature) is that biological and physical laws

---

[10]I am indebted to Mr. Harold Gans for suggesting this.

are suspended. One of the ways in which *HaShem* conceals his presence in the world is through action in a regular manner. When he acts in a manner we can predict, we have the option of claiming that a "physical law" is in effect, and we can remove God from the picture. But *HaShem* is not bound by physical law.[11] The scientific method can only discover regularities in the behavior of the world and predict what will happen when these regularities, which we call "laws," are in effect. If we have been told that these laws have been suspended, all bets are off.

So the fissioning of *Adam* into a man and a woman does not constitute a contradiction of biology. But is there any scientific evidence that confirms this rather unusual event? In a somewhat indirect way, there is.

Within the nuclei of the cells of our bodies are long strands of DNA that carry the words called genes. We get some of these strands from our fathers and some from our mothers. Because each parent contributes half, the various genes sort of mix and match and produce offspring who are genetically different from either parent. Yes, we might inherit mom's long fingers and dad's middle-age paunch (eventually), but we might also find that our dark-haired parents have given birth to a blond child. It is this shuffling of genetic structure found in sexual reproduction that gives such richness to our species.

So we get half of our genetic material from our mother and half from our father, correct? Wrong! It is only in the

---

[11]Nor, perhaps, are we. Rabbi Aryeh Kaplan comments that the uncertainty and unpredictability of the world of very small things is evidence of the incompleteness of physics. The universe is but an arena of action for free-willed creatures, meaning us. [Rabbi A. Kaplan, *The Handbook of Jewish Thought* (New York: Maznaim Publishing Company, 1979), section 3.33].

nucleus that half comes from each parent. There is, however, genetic material elsewhere in the cell. Within these little living organisms are structures called *mitochondria*, complex bits of organic material that are involved with the use of energy in the cell. The mitochondria have their own genetic material—strands of DNA—and all the mitochondrial DNA comes from the mother, none from the father. Because there is no shuffling from generation to generation, all variations in mitochondrial DNA are due to random mutations, and changes occur quite slowly.

In the mid-1980s, a trio of geneticists named Allan Wilson, Rebecca Cann, and Mark Stoneking began to study the mitochondrial DNA from people of different races living in widely different areas of the planet. They found remarkable similarities in their genetic material. The differences were so slight that it was very unlikely that all these people had descended through different evolutionary paths. The researchers felt that they had proved that all living human beings descend from a single woman who lived in Africa thousands of years ago. They laughingly called her "Eve," and the discovery became so famous that it even made the cover of *Newsweek* magazine.[12]

Since this time, the discovery has become controversial. Although most geneticists agree with the hypothesis, some paleontologists claim that the fossil record is not consistent with it. The geneticists in turn say to the bone experts, "You just have not found enough of the bones to know for sure.[13]

The jury is still out, but since we are just thinking about creation, it is very tempting to speculate. The Torah tells us

---

[12]*Newsweek*, January 11, 1988.
[13]J. Shreeve, "Argument over a Woman," *Discover* 11:8 (August 1990): pp. 52–59.

that we all descend from one great-great—very great—grandmother (and grandfather). Although modern biology cannot prove this incontestably at this time, current theories are certainly consistent with what it says in Genesis. But modern biology can confirm only the existence of a single female ancestor. There is no genetic material in the cell that seems to come only from the father, and hence we cannot (at this time) tell anything scientifically about our male progenitor(s).

Interesting. In the Torah there is an asymmetry. *Chava* (Eve) is called the "mother of all flesh," but there is no similar statement about her husband being the father of all flesh. Mitochondrial DNA seems to show us that there was a mother of all flesh; there doesn't seem to be any way of verifying a father. Is the Torah making only those statements that we can verify?

We will have to wait a few years until the controversy dies down. Then, we may or may not know. Scientific proof of the surgery in the Garden? Not yet. But a glance at items 26 through 29 on the blackboard shows that there is certainly nothing inconsistent between Torah and science when it comes to the origins of man. Indeed, the accounts of both disciplines are matching quite nicely. But there is one *kashia* we have glossed over, and it is a big one. Discussion of this will have to wait until the next chapter.

23

The Bored Clerk
The Relativity of Time

The big bang. *Tohu VaVohu*. The origins of matter. Life. Man and Woman. We have seen some striking parallels between the Torah and scientific views. The two disciplines are coming closer and closer together as they must. But we are not there yet. There is a glaring *kashia* that remains.

This is an obvious problem that we have not mentioned so far, and it surely must have occurred to many readers. It is the question of time scale. The Torah tells us that the universe was created in six days and that the land animals and man were created on the sixth. Current scientific theories claim that the big bang occurred about fifteen billion years ago, that the earth is about five billion years old, and that life took millions of years to evolve. How could the Torah and scientific descriptions both be true?

OK, we do have some problems in understanding the plain meaning of the Torah text. What does the word "day" mean? Usually, the word refers to the interval between one sunset and the next, but the sun was not created until the fourth "day." What did this term mean before then? Could it have referred to some longer period of time? From the written text alone, it is hard to tell.

There are also some internal problems. God brought all the animals of the world before *Adam* to see what he would name them. Now, how long does it take to visit the zoo? We can spend a good Sunday there and still not see all the animals. How long would it take for *Adam* to see and name all the thousands of animals that exist? All this happened on a very busy day during which *Adam* was created, split into a man and a woman, and got into trouble by enjoying a meal he shouldn't have. There are only 86,400 seconds in a day. How did he fit it all in?

It is very tempting to say that the word "day" here means something else. But, when we look into the classical commentaries on the Torah, we see that most sources seem to take "day" to mean an interval equal to our modern twenty-four-hour period. We are left with several small internal problems and one big *kashia* between Torah and science.

One source says billions of years; the other says six days. They can't both be right—can they? Let's digress and recall the conclusion of this book. Some apparent contradictions can be resolved; some cannot, within the limitations of our present knowledge. What about this one?

If we were thinking about this a hundred years ago, we would be stuck. All we could do would be to say "we do not yet know how to resolve this *kashia*. Perhaps our scientific knowledge is incomplete. Perhaps there is something about Torah that we are missing. We have confidence that someday the *kashia* will be resolved, but we are just not there yet." This would have been an honest statement, but it certainly would have been disappointing. After all, this is such a major *kashia*.

It would seem almost impossible to resolve, no matter how radically we modified our view of the world.

But this is not the nineteenth century—it is the end of the twentieth—and that changes everything. In the interim, something happened. The year was 1905, and in Bern, Switzerland, there was a young patent clerk who was very bored with his job. He was so bored that he used his free time to solve the three outstanding problems in nineteenth century physics. The young clerk's name was Albert Einstein.

The first problem he solved was called the Brownian motion. When small objects, such as pollen grains, are viewed under a microscope, they don't stand still. Instead, they keep jiggling around. Why? Einstein realized that pollen grains are so small that the molecules of water or air surrounding them, which are in constant motion, bump into them and cause them to jiggle. This explanation helped establish the reality of atoms and molecules.

The second problem Einstein solved was that of the *photoelectric effect*. When light shines on a metal, it causes electrons to be ejected from the surface of the material. These electrons come off with an energy that can be measured. If light were a wave phenomenon, we would expect that increasing the strength of the light would make the electrons pop out with greater energy. This is not the case. Instead, increasing the intensity of the light causes *more* electrons to be emitted, but they are always emitted at the *same* energy. The only way to change the energy of individual electrons is to change color of the light. None of this made sense to physicists.

Einstein suggested that light energy could not be absorbed by the metal in arbitrarily small amounts. Rather, it could only be absorbed in fixed packages, which he called quanta. The size of the quanta depended only on the color of the light. Changing the intensity of the light didn't change the size of the quanta, so the electrons still came off at the same energy.

Why would light act as if it were composed of little packets of energy? Perhaps it was made up of little particles after all, just as Newton had said. This led to the *kashia* between the wave and particle theories that we discussed in Chapter 5. Resolution of this *kashia* led to the quantum theory. Einstein's explanation of the photoelectric effect was so important that it earned him a Noble Prize.

But, it was Einstein's solution of the third problem that brought him his greatest fame. He was able to explain a very enigmatic observation by two American physicists named Albert Michelson and Edward Morely. To understand what Michelson and Morely were doing, let us first consider Mark Spitz, who is an Olympic swimmer.

Mark is fast. He can swim at about four miles per hour, which is equivalent to a good brisk walk. Let us suppose that he dives into a river in which the water is flowing at five miles per hour. If he floats on his back, he will be carried downstream at this speed. Suppose he wishes to swim back upstream, can he do it? Since the current is taking him downstream at five miles per hour, he would have to be able to swim faster than this to be able to fight the current and get back to where he started. He can't swim *this* fast, so he is not going to make it. On the other hand, if he tries swimming downstream, with the current, his four miles per hour will add to the five miles per hour of the water, and he will end up zipping down the river at nine miles per hour. Now he's really moving!

Michelson and Morely thought that they could do the same thing with light. Light moves pretty fast, at about

186,282 miles per second. The earth doesn't move quite this fast but, as it revolves around the sun, it does travel at a still-respectable 18 miles per second. As we mentioned in Chapter 17, nineteenth-century physicists thought that all space was filled with a mysterious substance called ether and that light waves were vibrations of the ether. If this were true, the earth would feel an *ether wind* due to its motion through the substance, and light would move faster or slower than 186,282 miles per second depending upon whether it was going with or against the current (the ether wind). To their great surprise (and everybody else's), there was no such effect. Light traveled at the same speed no matter which direction it went.

The only person who seemed to be able make sense of this was the young patent clerk. He published a paper called *Zur Elektrodynamik bewegter Koerper* ("On the electrodynamics of moving bodies")[1] with a new theory that came to be called *relativity*. The theory of relativity goes something like this: there is no ether wind; there isn't even any ether. That's the easy part. Now for the hard part. The speed of light is always the same no matter how or where we measure it. Whether we are moving toward or away from the light source makes no difference. We always get the same answer.

Now this did not seem to make sense. Suppose a flashlight is moving toward us at 90 percent of the speed of light, throwing its photons (little particles of light) in our direction. Wouldn't the photons come at us at a greater speed than if the flashlight were not moving at all?

---

[1] A. Einstein, "On the Electrodynamics of Moving Bodies," in *The Principle of Relativity*, trans. W. Perrett and G. B. Jeffery (n.p.: Dover Publications, n.d.), pp. 35–65.

Einstein said no. The speed of light is always the same. How could this be? His answer was the truly difficult and deep part of the theory, and it required a radical change in the way physicist's view the world.

To determine the speed of anything we must measure both distance and time. We need rulers and clocks. Einstein realized that rulers and clocks do not behave as nicely as most of us think.

When we watch a clock and determine that an hour has passed, we intuitively believe that everyone would agree with this; an hour is an hour. That is, we feel that time is an *absolute* quantity, one whose value is the same no matter how we measure it. An example of an absolute quantity is the floor number of a hotel room. If we are on the seventh floor, everyone agrees with this, even those on higher or lower floors. The seventh floor is the seventh floor. But is the seventh floor in the upward or downward direction? Marvin, on the second floor, claims the seventh floor is upward, while Judy, on the ninth floor, claims it is downward. Can they both be correct? Of course, because upward or downward are not absolute quantities, they are *relative* quantities. The seventh floor is upward relative to Marvin on the second floor and downward relative to Judy on the ninth.

Einstein was able to show that distance and time are relative quantities, not absolute ones. That is, one person might read his clock and determine that an hour has passed, and another may look at his clock and claim that a minute has passed, and *they are both correct*! Time has meaning only *relative* to the manner in which it is measured. The same thing happens with distance. Bill might measure something and find it to be one inch long, while Alice finds it to be two inches. Time and space are not absolute concepts. They are only meaningful *relative* to an observer. Because of the strange behavior of time and space, whenever we determine the speed of light, we come up with the same answer.

The implications of this are astounding. Let us consider a pair of twin brothers: Reuven and Shimon, both 40 years old. Reuven is a homebody. He has a nice job and family and lives in Brooklyn. But Shimon has a wanderlust, and he has just accepted a job as an astronaut. The pay is good, and he will have a lot of time to study Talmud since he will be going to a star that is twenty light years away. (That is, it is so far that it takes twenty light years to get there.) The big day comes, and Reuven wishes Shimon good-bye. Off he goes at nearly the speed of light. It takes Shimon a little more than twenty years to reach the star, a month to explore it, and a bit more than twenty years to return. Finally, Reuven goes out to meet the brother he has not seen in over forty years. Reuven is not as agile as he was when Shimon left, since he is now a man of eighty, and he is quite surprised to see his brother zip down the gangway still a young man, for he has aged only a bit more than one month!

What has happened? For Reuven, forty years have passed, but Shimon has been moving very fast, and the theory of relativity tells us that when things move fast, time slows down for them. Only about a month has gone by for Shimon. More technically, *relative* to Reuven's *reference frame* forty years have elapsed, but *relative* to Shimon's *reference frame*, only a month has passed.

As amazing as this must seem, it is more than a speculative theory. It has been tested carefully and can be observed in the laboratory. There are subatomic particles called mesons that are produced in the upper atmosphere of the earth. We know that mesons can only live a short time before they decay into other particles. They just don't have a long enough lifetime to make it down to the surface of the earth. Nevertheless, we can detect mesons at the earth's surface. Why? Because mesons move very fast, and in the meson's moving reference frame only a short time has passed.

We have mentioned that Einstein produced two theories of relativity. The 1905 theory is called special relativity and deals with uniformly moving objects. Later, in 1917, he extended this to a theory of general relativity, one that also explained accelerating objects and gravity.

The special theory tells us that time is affected by speed. The general theory predicts that time is also affected by gravity. Time flows more slowly near the floor of a room than it does near the ceiling. The effect is so small that we never notice it in our lifetimes, but it is possible to do experiments with extremely accurate clocks that confirm Einstein's theory.

In regions of space in which gravity is very strong, time can slow down quite a bit or even stop! One such place is near a strange object called a *black hole*. A black hole is a star or other conglomeration of matter that is very dense and that produces an immensely strong gravitational field. This field is so strong that nothing, not even light, can escape from the black hole. Surrounding the black hole is a region of space called the *event horizon*. If anything gets closer than the event horizon, it can never escape from the black hole. Never. Don't try it!

Recently, astronomers have used the Hubble telescope to confirm the existence of a black hole in a nearby galaxy. Some astronomers believe that there is a black hole at the center of our galaxy, and there may be others within it. Indeed, there are some physicists with an even stranger hypothesis. They speculate that the physical vacuum, the underlying *Tohu VaVohu* of the universe, gives rise to virtual black holes, small ones appear and disappear in the space between atoms. If this is true, there are microscopic places within all matter in which time slows down or even stops.

It is not only intense gravitational fields that do funny things to time. Empty space can also pay tricks. In Einstein's theory, space is not passive. It is not just an arena in which things happen. Instead, space is active and dynamic. The fabric of space itself can become distorted, and when this happens time slows down. This may sound like science fiction or poetry, but it is not. It is serious science, and the rather complicated equations of general relativity give it definite meaning.

So there are many ways in which time can flow at differing rates. The principal point is the following: two people can claim that different intervals of time have passed *and they can both be correct*! One can say that an hour has passed, another a minute. One can observe that six days have passed, and the other can claim that fifteen billion years have elapsed, *and they can both be right*. Strange as it may seem, there is no intrinsic contradiction between the time scale given in Genesis and that claimed by the scientists!

Of course, just showing that there is no intrinsic contradiction is not enough. We would really like a more detailed explanation of how it all works out. To do this, we must identify the different reference frames used by the Torah and the physicists, and show how six days elapse in one while fifteen billion years elapse in the other.

We cannot yet do this with any certainty, but there is one fascinating possibility that tempts us to speculate. When we look into *Mishnah*, it tells us the dimensions of the Holy of Holies, the innermost chamber of the Jerusalem Temple. We are told that the distance from each side of the ark of the covenant to the adjacent wall was 10 cubits, and that the chamber was 20 cubits across. But—just a minute. How could this be? Didn't the ark take up any space itself? There seems to be something wrong with the arithmetic.

This problem exists only if the space within the chamber is space as we usually experience it. Suppose the space within it were somehow distorted (physicists would say bent). Then, the dimensions could be distorted also. It would explain the funny measurements.

If the space were distorted, so would be time. It would move much more slowly than time external to the chamber. Perhaps we have found the reference frame of the Torah, the place in which six days elapsed while the universe as a whole aged fifteen billion years. The Temple was built on a rock called the *Even Shesiah*, the foundation stone. Tradition tells us that this was the first solid piece of the earth to be created. What more fitting reference frame could be found to tell the story of creation?

We ended Chapter 19 with a question. When we look at the sequence of events given in the Torah, there is one thing that seems to be out of sequence: the creation of the earth. It occurs too early in the process. After all, the planets were created much later, along with the sun and the moon, weren't they?

But if the *Even Shesiah* was the first thing to be created, perhaps the earth did come first. Perhaps God first created the earth, and then the other heavenly bodies. Perhaps, in the distorted space near the *Even Shesiah* only six days elapsed while the rest of the universe aged several billion years. Wouldn't this make it all fit together nicely?

Perhaps. Or—perhaps not. It is really just speculation, and pretty wild speculation at that. But we do it for a purpose: to show how little we really understand. What we do know for certain is this: there is no intrinsic contradiction between science and the Torah when it comes to time scale. We haven't worked it all out yet, but we are certainly a lot closer to resolving the *kashia* than we were a hundred years ago!

24

Putting It All Together
The Story of Creation

We have come a long way—from the first moments of creation to the mother and father of us all. We have seen amazing similarities between the Torah and scientific viewpoints. Now would be a good time to put it all together and see how things fit, and where they don't—yet.

To make it easier to follow, it will be helpful to consult the blackboard as we go along. The right side of it is all filled in now, and it can be found in Figure 24–1. We will use numbers in parentheses to refer to lines on the blackboard. Although the Torah tells the story of creation twice, we will be interweaving the events of the two versions into a single narrative, and this will be apparent from the line numbers. Let's start at the beginning.

There was a moment of creation (1). At this instant, God willed the universe to come into being from nothing. This moment was the origin of time and space, so it is meaningless to ask what happened before creation. The event cannot be explained by the laws of physics. Cosmogonists refer

TORAH	SCIENCE
1. Moment of creation —something from nothing	1. The big bang
2. Continuously renewed creation	2. Continual creation and destruction of particles
3. Chaotic nothingness	3. Physical and false vacuums
4. Creation through 10 utterances	4. Need for "classical observer"
5. Light	5. Light
6. Some light hidden away	6. Cosmic background radiation
7. Six twenty-four hour periods	7. Relativity of time
8. Concealment of single force	8. Concealment of single force
9. Fluid/vapor gathered together	9. Cosmic gas clouds
10. Force towards one place	10. Gravitational collapse

Figure 24–1, part 1: The Blackboard

to this instantaneous explosion of time and space as the "big bang."

Initially, the universe consisted of nothing but empty space. The nothingness was chaotic, *Tohu VaVohu*, a state physicists call the false vacuum (3). God had created laws of nature so that the world would operate in a predictable fashion. In this manner, He hid His existence from us. One consequence of these laws is that universal nothingness became unstable, and out of it came something: a very hot and very intense field of radiation, what we call light (5). For this to

TORAH	SCIENCE
11. First fluid bodies	11. Solar and planetary nebulae
12. First solid matter	12. Planets
13. Genetic material	13. Panspermia?
14. The heavenly bodies	14. Stars
15. The sun and moon	15. The sun and moon
16. Animate life: birds and fish	16. Animate life: fish, birds?
	17. Animate life
17. Animal soul	18. Land animals, birds?
18. Land animals	19. Homo sapiens
19. Home sapiens	20. Programmed cell reproductions
20. Long life span	
21. First human soul	21. First human soul

Figure 24–1, part 2: The Blackboard

happen, it had to be observed by an intelligence, and that intelligence was God, who took cognizance of the universe through formulation of His ten utterances (4).

Although the particles of radiation and all other particles that later appeared in the universe appear to be continuously present, this is an illusion. When particles interact with each other, what actually happens is that the original particles disappear and new ones are created in their place. In this manner, the creation is continually renewed by God (2).

As the universe expanded, it cooled. Today, some of the original light of creation remains, filling all of space with the very cold cosmic background radiation (6). Until recently, this

TORAH	SCIENCE
22. Mist waters earth	22. Condensation of water
23. Man from dust of earth	23. Life from clay?
24. Vegetation	24. Plant life
25. Animals from dust of earth	25. First animals, also from clay?
26. Man looks for mate	26. Other homo sapiens
27. Man cannot find partner	27. Only being with a soul
28. Hermaphrodite split apart	28. Miracle outside laws of nature
29. Female is mother of all flesh	29. Mitochondrial DNA

Figure 24-1, part 3: The Blackboard

light was concealed, and it is only in the last few years that it has been revealed to man. At the present time, we can observe this light, but we have no use for it.

There is actually only one force that operates in the universe. Yet, as the universe developed, it went through a series of condensations in which this force became deeply concealed so that those of us who are made of ordinary matter experience the illusion of many forces (8). This is yet another way in which God hides His control of the world. As a result of these condensations, the primal radiation gave forth more substantial particles of matter, so that space became filled with a diffuse soup of radiation and matter.

The universe was now about a million years old. Yet, since time is a relative rather than an absolute quantity, to other frames of reference only one day had elapsed (7). It continued to cool, and the matter began to gather into immense clouds of gas (9). The force of gravity caused these clouds to collapse inward to form giant nebulae (10). The nebulae themselves collapsed to form stars and planets (11, 12). These stars lived, died, and exploded, spewing their contents into space as gas clouds that then condensed and started the process all over again.

Somewhere, somehow, God created genetic material (13). Perhaps it was in space. At present, we just don't know. What we do know is that about ten billion years after the big bang (about four days later, according to other reference frames) a collapsing gas cloud gave birth to a smallish yellow star we call the sun, along with its planets and their moons (14, 15).

The earth was initially barren, but as it cooled water condensed to form clouds that gave forth rain to water the earth (22). This water caused genetic material on the earth's surface to blossom into the first simple life. Perhaps some of this genetic material originated in space and fell to earth (13), perhaps life began in clay upon the earth (23, 25), or perhaps both. In any case, simple life appeared followed by vegetation (24). Fish and birds (16) and land animals (18) then appeared, although the animate souls of these beings were a new creation by God (17). This process happened in a manner consistent with the laws of biology, perhaps through a slightly modified version of Darwin's theory of evolution by mutation and natural selection. While biologists would call this a random process, because they cannot discern a purpose in the "accidental" mutations, the Torah tells us that there are no accidents in life, and so all evolution actually was guided by God toward a particular end: the creation of man.

Toward this end, God arranged the laws of nature so that mammals eventually evolved into a being we call Homo sapiens (19), biological man. But the first members of these species did not have human souls. After 974 generations—which might have been quite long compared to modern life spans (20)—God took one of these beings, a hermaphrodite, and placed within him/her the first true human soul (21).

This spiritually advanced being could not find a companion among the other biological beings who lacked human souls (26, 27), so God caused a miracle to happen that suspended the laws of nature. This being fell into a deep sleep and was fissioned into separate male and female persons (28) who are the ancestors of the present human race, the woman being the mother of us all (29), and the man the father.

And so our story comes to an end—or rather, it begins.

25

. . . Or Maybe Not
Alternative Approaches

The story in the previous chapter is so clear, so nicely put together, so understandable. It's too bad there are problems with it. Not severe problems, not problems that threaten the relationship between science and Torah, but *kashias* nevertheless that must be resolved someday, even though we cannot do so with our present level of knowledge.

For example: the Torah reports that light (radiation) was created on the first day, and that the *Rekiah*, which we identify with the *tzimtzum*, occurred on the second. We have previously observed that we cannot always assume that the Torah is written in chronological order, but when Genesis assigns one event to the first day and another to the second it is difficult to imagine that they occurred in a different sequence. The scientific viewpoint is a bit cloudy here. We have previously mentioned that there is some uncertainty in just how we define "light." When the Torah refers to light, is it talking about the sea of massless particles that existed before the single universal force was hidden? If so, we have perfect consistency between the scientific and Torah views. On the other hand, does "light" refer only to the melange of radiation that existed later in creation? If so, we have a *kashia* that must be resolved.

What about the creation of the heavenly bodies? Looking at our blackboard, the Torah is a little unclear on the order in which these bodies were created. We have identified the appearance of the first solid matter (*yabasha*) as the appearance of the planets (12). Only later did the stars appear (14). This is not really in consonance with the order of modern cosmogony. But there are many uncertainties here. We have seen that stars and planets were continually formed and dissolved many times in the history of the universe. Which incarnation of stars and planets is being spoken of in Genesis? Also, our understanding of the Torah text is a bit uncertain. Are we correct in identifying the solid matter as the planets and the heavenly bodies as the stars? Perhaps the *yabasha* represents bits of solid matter that formed in space before gravitational collapse produced the first stars. Our knowledge here is still somewhat soft. There is more work we must do.

Somewhere on the third day, God formed the first genetic material. This is consistent with the theory of an extraterrestrial origin for life (panspermia), but what about the other theories of life's origins? Those theories that assume life originated in clay are consistent with the second version of the creation story, but are they with the first? Although the situation is far from clear, we do observe that the Torah itself seems to present a *kashia* between the first and second versions of the story. So, the problem doesn't really lie in a contradiction between science and Genesis. Rather, it is a difficulty in understanding the plain text meaning of the Torah itself.

When it comes to the evolutionary sequence of life forms, there is another problem. Genesis seems to say that birds (*oafos*) evolved before land animals (birds on the fifth day, animals on the sixth). Biologists currently disagree with this order. Perhaps the Hebrew word *oaf* refers to other flying creatures, such as insects? Or perhaps there is a problem in the scientific picture, and birds did come first. It is just too early to tell.

None of these problems are show-stoppers. None of them cause us to abandon our suggested picture of "how it all fits together." They represent remaining *kashias* that are yet to be worked out as we continue to think about creation. But suppose someone does come up with a *kashia* that is so threatening that it severely damages our picture. Does this mean that we must finally give up the idea that science and Torah are consistent?

Of course not. It would only mean that we have been thinking about things in the wrong way. It would mean that we must reframe our thinking in a different manner, and we would learn much in the process. Indeed, there have been many great thinkers who have approached the entire subject in far different ways, and since any one of them might be correct while we might be wrong, it is important to review what they have said.

One approach taken by a number of scholars is to regard the opening sections of Genesis as allegorical. In recognition of the purpose of Torah—to reach us how to live rather than to teach us science—this viewpoint emphasizes the moral lessons in Genesis and deemphasizes or ignores the question of how much is literally true. Such an approach is hinted at by Rav Kook when he says:

It makes no difference for us if in truth there was in the world a Garden of Eden . . . .[1]

While this approach instantaneously solves all the *kashias* we are likely to run into (we just ignore the literal words of the Torah), a look at the overwhelming majority of classical commentators on the Torah seems to indicate that they *do* take the words of Genesis to be literally true. This is not really a satisfying way to deal with the question.

Another approach taken by some is to note that we cannot observe the past. All scientific evidence—fossils, light that appears to have been traveling billions of years from distant galaxies, cosmic background radiation—consists of phenomena that exist today. We assume that the laws of nature have always been as they are now, and then we theorize about the past. But suppose God is fooling us. Suppose He created the universe instantaneously just as we see it: with fossils in the ground, with light falling on the earth as if it had come from distant stars, and with the cosmic background radiation already in place. We would have no way of knowing that this trick had been played on us, and as good scientists we would construct a story that was intelligent, well reasoned, and false.

There is no way that anyone can prove that this did not happen. Indeed, we cannot prove that the world was not created just one minute ago, even though we all remember things happening yesterday. Maybe we were all created sixty seconds ago with all our false memories stored within our brains.

This would certainly be an effective way for God to hide His existence from us. But, although He might hide from us,

---

[1]T. Feldman, *Rav A. Y. Kook, Selected Letters* (Ma'aleh Adumim, Israel: Ma'aliot Publications, 1986), p. 12.

would He create a world designated to lead us astray? We cannot second guess God, but it just does not seem fair.[2]

A third, and far more satisfying approach, is to observe that scientific theories are only tentative attempts at learning the truth, that they cannot ever be totally proven, and that gross errors can and do occur. Perhaps the entire picture of a slowly evolving universe fifteen billion years old is completely incorrect. Rabbi Menachem Mendel Schneersohn has pointed out that theories of the past are based upon extrapolation, a method of reasoning subject to significant error.[3] Therefore, our conclusions as to the origins of life and the universe could be wrong.

For example, the age of the universe is only a theoretical construct. It has been derived from the expansion of the universe, which in turn is inferred from the red shifts of distant galaxies. What if this red shift has a different cause, undiscovered as of now? Then our calculations would be totally wrong. Indeed there are some reputable astronomers who speculate that this is the case. As another example, let us consider the age of the earth. This is derived from measurements of the ratio of uranium and lead found in rocks. But the calculation

---

[2] A variation on this idea has been suggested by David Shulman. He speculates that perhaps God created the world several millennia ago and, being outside of time, also then caused time to flow forwards towards our present day and backward from the moment of creation for billions of years to the origins of the big bang. This is an ingenious suggestion, but it is one that cannot be examined by the scientific method.

[3] Rabbi M. M. Schneerson, "A Letter on Science and Judaism," in *Challenge*, ed. Aryeh Carmell and Cyril Domb (New York: Feldheim, 1976), pp. 142–149.

requires that we know how much of each were present originally, and these quantities are not really known, only assumed.

The ages of fossils are also uncertain. Some are inferred
from the sequence of fossil layers—the deeper, the older. In
some cases, however, the layers occur with the "younger"
fossils underneath the "older" ones. Geologists explain these
by assuming that the rock layers have flipped over. Other fossil
ages are derived from radioactive carbon dating. Carbon dating requires us to make assumptions about the initial amounts
of radioactive material that were present in the organism, again
an assumption rather than an observation.

None of this is intrinsically bad science. Scientists have
taken a large amount of indirect evidence and constructed a
beautiful and (for the most part) consistent story that agrees
with the evidence. This is the best they can do, and it is actually pretty good, but it is not the same as proof.

There is another, more subtle problem with scientific
knowledge. In Chapter 3 we discussed the manner in which
scientists decide which theories are accepted and which rejected. We described the process of peer review. We saw that
the use of expert referees to decide which scientific papers are
published was a practical necessity, but that it could lead to
the suppression of unpopular theories or results.

Again, we wish to stress that this problem is not fatal to
science. Most researchers are intelligent, conscientious, and
honest. They, too, are aware of the problems of peer review.
But they just don't know of any better way of deciding who
gets into print and who doesn't. It is an attempt at fairness.
It fails sometimes—it is imperfect—but no one has ever been
able to think up a better way.

Thus, it is quite possible that there are serious errors with
the current scientific picture of the universe's history. Yet we
must admit that researchers still, in general, do a good job.

We cannot casually throw out theories that cause *kashias*. We must work with the world as we understand it at the moment. As we have said previously, this book is for those who *do* take science seriously as they think about creation.

Finally, we must mention the approach of the great scholar Rabbi Yisroel Lipshitz, author of the *Tiferes Yisroel*. He has observed that *Kabbalah* (the esoteric part of the Torah) tells us that God created and destroyed many worlds before this one. He posits that the evidence of previous eras, engraved in the layers of fossils around the world, is the remnant of these previous worlds. This is also the position taken by Rav Kook.[4] The previous worlds could be a reference to earlier eras on this earth, or perhaps to the previous planets and stars that were created and destroyed before our solar system. This is a controversial position among some scholars, who assume that when the Torah speaks of previous worlds that God created and destroyed, it means that they were destroyed without a trace.

All the above positions make sense; any one of them could be the truth. But we have taken a more direct approach. We have chosen to take the current state of science seriously and to think about it in the context of Torah. We have found that there is no need to go as far as the above viewpoints do. Science and Torah are currently amazingly consistent, and they seem to be drawing closer every day.

Do we have the whole truth? Of course not. Not yet. But at least we have a way of thinking, and think we shall as we continue to delight in our examination of creation.

---

[4]T. Feldman, *Rav A. Y. Kook, Selected Letters* (Ma'aleh Adumim, Israel: Ma'aliot Publications, 1986), p. 5.

26

The Bottom Line (Again)
In Conclusion

We can be fairly sure that someone reading this book a century from now will be tempted to laugh at some of our ideas, realizing that they are based upon outmoded scientific theories. If our bottom line were, "This is the way science and Torah fit together; this is the way it all works out," he would be right. But this is not where we leave our study.

We know that tomorrow the next generation of telescopes may tell us that our ideas of the cosmos need massive reworking. The behavior of a subatomic particle may violate the most deeply held theories of physics, and Noble Prizes will be awarded to a new generation of young physicists who will again "work things out." Doubtlessly some of them will think that they have finally achieved the ultimate understanding, but we know otherwise. We know that *HaShem* has created a universe that can be understood only approximately, both in its physical and spiritual dimensions.

But this does not depress us. After all, a small child is not depressed by his lack of knowledge. In his innocence he

289

sees the mystery of the world as a wonder rather than a threat. By learning how to *think about creation*, by concentrating upon the process of gaining understanding rather than expecting to reach the total knowledge that is forever beyond our grasp, we can come to appreciate the magnificent depths of the Creator. By seeking to find Him where He has hidden Himself, we can experience His presence through the wonder of it all.

Isn't it going to be exciting?

Glossary

adam a human being

adamah earth

amino acid one of the building blocks of a protein

androgynus a hermaphrodite

astronomical unit the average distance from the earth to the sun

bara created

baryon a type of subatomic particle

base one of the building blocks of a DNA molecule

baseline the distance the observer moves during a parallax measurement

big bang the explosion that gave rise to the universe

black hole	a concentrated massive body whose gravity is so strong that light cannot escape from it
boson	a type of subatomic particle
Cepheid variable	a type of star with time-varying brightness
chait	sin, transgression, missing the mark (see Chapter 8)
chaos	the mathematical theory that explains systems in which very small disturbances can have large effects
charge conjugation	the interchange of positive and negative charges
Chava	Eve
chessed	loving-kindness
chidush	an original way of looking at Torah
classical physics	physics as it existed before quantum theory
cosmic background radiation	the cold light left over from creation
cosmic inflation	a theory of the initial expansion of the universe
cosmogony	the study of the origins of the physical universe

cosmology   the study of the structure of the universe

creation and   in quantum field theory, agents of
  destruction   creation and destruction of particles
  operators

cybernetics   the study of the similarity between the
                     brain and the computer

dinar   a coin equal in value to 192 *perutas*

DNA   the long molecules that store genetic in-
                     formation in the nucleus of a cell

electromagnetism   the single force that underlies electricity
                     and magnetism

Elokim   God

electroweak force   the single force underlying the electro-
                     magnetic and weak forces

entropy   a measure of the disorder in a physical
                     system

ether   a substance nineteenth-century physicists
                     believed filled all space

even shesiah   the foundation stone, the first part of the
                     earth to be created

ezer knegdo   a helper against him

false vacuum   the initial "less than nothingness" of the
                     universe

field   something that spreads out and fills
                     space

first law of thermodynamics	the conservation of energy
gadol	large or great; applied to a Torah scholar
gedanken experiment	an experiment performed in the mind
genetic code	the code that associates base sequences in the DNA molecule with amino acids in a protein
grand unified theory	a theory that unites the strong and electroweak forces
g'vurah	strength and justice
halacha	Jewish law
halachic	legal
haolam	the universe
HaShem	literally: "the name," the unpronounced four letter name of God
havdalah	the ceremony at the conclusion of *Shabbos*
ish	a male human being
isha	woman
kabbalah	the hidden esoteric material of the oral tradition
kashia	an apparent logical contradiction
kilayim	forbidden mixtures

klal uprat	a generality followed by a specification
klal sheleacharav ma'aseh	a generalization followed by a particular event
kosher	fit, proper, permissable to eat
lepton	a type of subatomic particle
light year	the distance light moves in one year
ma'aser ani	the tithe for the poor
matanos aniim	gifts to the poor
mayim	water, but also fluid or vapor (see Chapter 7)
melacha	creative work
melanin	a brown pigment found in the skin
meson	a type of subatomic particle
Midrash	the homeletical part of the oral tradition
Mishnah	the core text of the Talmud
mitochondria	small structures within a living cell
mitzvah	a commandment, a good deed, an act desired by God
muon	a type of subatomic particle
mutation	a random change in the genetic material of an organism
nebula	plural: nebulae. An immense cloud of gas in interstellar space

nefesh	the animal soul or life force
nefesh chaya	a living being
neutrino	a type of subatomic particle
nishmas chaim	the human soul
oaf	bird
oseh	completed
parallax	the apparent shift in position of a distant object caused by the observer's sideways movement
particle accelerator	a device that accelerates subatomic particles to very high velocities
peruta	a coin of small value
photon	a particle of light
physical vacuum	empty space, with the laws of physics in effect
plasma	an ionized gas, such as a flame
polypeptide chain	the chain of amino acids that makes up a protein
prat uklal	a specification followed by a generalization
pshat	the literal meaning of the biblical text
quantum	plural: quanta; a small, indivisible amount of energy

quantum field theory	the theory of subatomic particles and their interactions
quantum theory	the laws of physics that apply to very small things
quark	a type of subatomic particle
quasar	a quasi-stellar radio source
Rambam	Maimonides, twelfth-century philosopher and commentator
Ramban	Nachmanides, thirteenth-century Bible commentator
Rashi	eleventh-century Bible and Talmud commentator
rekiah	the firmament (see Chapter 7)
second law of thermodynamics	the law that the entropy (disorder) of the universe increases with time
seder	the festive ceremonial meal of the Passover holiday
Sforno	Italian Bible commentator, 1475–1550
Shabbos	the Sabbath
shichcha	the law that produce forgotten in the harvest must be left for the poor
Sh'ma	the twice-daily declaration of God's unity
shmitta	the Sabbatical year

spacetime	the four-dimensional extent of space and time
strong force	the short-range force that holds atomic nuclei together
Talmud	the written embodiment of the oral tradition
tanai	a disagreement between two rabbis at the time of the *Mishnah*
theory of everything	a theory that unifies all the forces of physics
thermodynamics	the science of heat
tohu vavohu	chaotic nothingness (see Chapter 7)
Torah	The Law, The Five Books of Moses
t'shuvah	repentance, return
tzimtzum	the kabbalistic term for God's contraction and concealment of His essence (see Chapter 7)
uncertainty principle	the basis of quantum theory: that physical quantities cannot be known with complete certainty
warfarin	a type of rat poison
wavelength	the distance between adjacent crests of a wave
weak force	the short-range force that causes certain types of radioactive decay

worldline	the path of an object in spacetime
yabasha	solid material
yatzar	formed
yetzer harah	the evil urge
yetzirah	formation
yesh meayin	creation of something from nothing
yetzer hatov	the urge toward good
yetzer hara	the internal urge to do something wrong
Yom Kippur	the Day of Atonement
Zohar	the primary written text of Jewish mysticism

Index

About the Author

Andrew Goldfinger was born in New York City and has a Ph.D. in theoretical physics from Brandeis University, and an M.S.E. in counseling from The Johns Hopkins University. Dr. Goldfinger currently serves as Assistant Supervisor of the Space Department's Mission Concept and Analysis Group at The Johns Hopkins University Applied Physics Laboratory. A frequent international lecturer for the Aish HaTorah Rabbinical College, Dr. Goldfinger resides with his wife, Shana, in Baltimore. They have three children.